ENGLISH
BLIND-STAMPED
BINDINGS

THE SANDARS LECTURES
1949

ENGLISH BLIND-STAMPED BINDINGS

BY

J. BASIL OLDHAM

M.A. (OXON.)

LIBRARIAN AND LATE ASSISTANT MASTER, SHREWSBURY SCHOOL;
SANDARS READER IN BIBLIOGRAPHY IN THE
UNIVERSITY OF CAMBRIDGE, 1949

'It is a golden rule, which I try to follow, to put every fact which is opposed to one's preconceived opinion in the strongest light.'

CHARLES DARWIN: *letter to* JOHN SCOTT *2 July 1864*

CAMBRIDGE
AT THE UNIVERSITY PRESS
1952

CAMBRIDGE UNIVERSITY PRESS
Cambridge, New York, Melbourne, Madrid, Cape Town, Singapore,
São Paulo, Delhi, Dubai, Tokyo

Cambridge University Press
The Edinburgh Building, Cambridge CB2 8RU, UK

Published in the United States of America by Cambridge University Press, New York

www.cambridge.org
Information on this title: www.cambridge.org/9780521136648

First published 1952
This digitally printed version 2010

A catalogue record for this publication is available from the British Library

ISBN 978-0-521-05861-2 Hardback
ISBN 978-0-521-13664-8 Paperback

IN PIAM MEMORIAM

GALFREDI DUDLEY HOBSON

QUI HIS STUDIIS INGENIUM SUUM ITA IMPERTIVIT
UT TIRONIBUS CONSILIO ERUDITIS AUCTORITATE
COMITER OPITULATUS
OMNIBUS INTEGRITATIS EXEMPLUM
ADMIRABILE PRAESTITERIT

PREFACE

IT was with considerable hesitation that I undertook these lectures, since, unlike most of my predecessors, I am not a professed bibliographer, but a mere schoolmaster, whose primary interest is, and always will be, the study and care not of books, but of boys. And my diffidence was only increased when I found that it was exactly fifty years since Edward Gordon Duff delivered his first Sandars Lectures, and, with Weale, Gibson, Gray and Gottlieb, laid the foundations of the scientific study of bookbindings. The only Sandars Reader since then to lecture on this subject was Geoffrey Hobson in 1927, and I should like to record the immense loss his death this year has meant to bibliography, as well as the personal loss to myself of a friend who started my interest in bindings, and who never failed to show the sympathetic tolerance and kindly generosity that characterises the disinterested scholar in his attitude towards beginners.

I wish to express my grateful appreciation of the patience and kindness shown by innumerable librarians in England and Scotland, who have given me facilities not only for visiting, but, in a number of cases, revisiting several times, their libraries for the purpose of taking rubbings, and, too often, of filling gaps in my notes that I had left through carelessness on previous occasions. They even treated my importunate and frequent letters with a cheerful tolerance that I did not deserve, and in several instances were good enough to save me a special journey by taking for me individual rubbings themselves. I should add that it would not have been possible for me to collect the rubbings for the reproductions that follow from bindings in libraries all over the country, but for a generous grant from the Trustees of the Leverhulme Research Fellowships Fund.

I am also indebted in particular to Mr H. M. Nixon, of the British Museum, for helping me with many useful suggestions and comments in preparing these lectures for the press, and to Mr Neil Ker, of Magdalen College, Oxford, for bringing to my notice several rolls that I had missed and for other help. I owe much, too, to the Cambridge University Press, in the persons of Mr R. J. L. Kingsford and my former pupil, Mr B. Crutchley, for the trouble they have taken to accommodate themselves to my whims and fancies in the production of this book: any faults in the lay-out of the plates (and they are many) must be laid wholly to my charge, not to theirs.

I have gratefully to acknowledge the gracious permission of H.M. The King to reproduce rubbings of stamps on the binding of Henry VIII's *Assertio* in the Royal Library, Windsor Castle. I wish also to acknowledge with gratitude the permission given me to reproduce rubbings of other unrecorded tools from bindings in their respective libraries by the Archbishop of Canterbury, the Marquess of Bath, Major J. R. Abbey, Mr G. Brudenell, Mr A. Ehrman and Mr G. P. Mander; by the Trustees of the British Museum and the Council of the National Library of Wales; by the Syndics of Cambridge University Library and the Curators of the Bodleian; and by the authorities

PREFACE

responsible for the following libraries: the Folger Library, Washington; Edinburgh, Glasgow, Aberdeen, St Andrews and Durham Universities; Gonville and Caius, Christ's, Clare, Corpus Christi, St John's and Pembroke Colleges, Cambridge; All Souls, Brasenose, Christ Church, Corpus Christi, Jesus, Magdalen, Merton, New College, Oriel, Queen's, St John's, Trinity, Wadham and Worcester Colleges, Oxford; the Public Record Office; the College of Arms; the Inner Temple; the Fitzwilliam Museum; the Society of Antiquaries; the British and Foreign Bible Society; Westminster Abbey; York Minster; Durham, Ely, Gloucester, Hereford, Lincoln, Peterborough, Rochester and Worcester Cathedrals; Downside Abbey and Bristol Baptist College; the Cities of Bristol and Norwich; the Boroughs of Colchester, King's Lynn and Wisbech; Chetham's Library, Manchester; Eton, Winchester, Stonyhurst and Ushaw Colleges; Shrewsbury School, Christ's Hospital, St Albans School and Guildford Royal Grammar School; and the incumbents of St Leonards, Bridgnorth, of Chirbury, Salop, and of Desborough, Northants.

Finally, the inclusion of photographs of entire bindings has been made possible by the kindness of Glasgow University; Gonville and Caius and St John's Colleges, Cambridge; All Souls, Jesus, Merton and New Colleges, Oxford; Eton College, Christ's Hospital and Shrewsbury School; and for the drawing showing the different types of design used on English bindings, and for the illustrations of the Glossary, I am indebted to Mr J. M. Woodroffe, late of Shrewsbury School.

J. B. O.

December 1949

CONTENTS

LIST OF PLATES

ALL RUBBINGS HAVE BEEN REPRODUCED IN THE SIZE OF THE ORIGINAL, AND ALL
BINDINGS EXCEPT WHERE IT IS STATED THAT A REPRODUCTION IS REDUCED

ABBREVIATIONS

THE FOLLOWING ABBREVIATIONS ARE USED IN REFERENCES TO BOOKS.
FOR ABBREVIATIONS IN REFERENCES TO ROLLS, SEE P. 41.

DUFF, *Westminster*	E. Gordon Duff, *The Printers, Stationers, and Bookbinders of Westminster & London 1476–1535.*
GIBSON	Strickland Gibson, *Early Oxford Bindings.*
GOLDSCHIMDT	E. P. Goldschmidt, *Gothic and Renaissance Bookbindings.*
GRAY	G. J. Gray, *The Earlier Cambridge Stationers and Bookbinders.*
HAEBLER	K. Haebler, *Rollen- und Plattenstempel des XVI. Jahrhunderts.*
HOBSON, *Cambridge*	G. D. Hobson, *Bindings in Cambridge Libraries.*
HOBSON, *Before 1500*	G. D. Hobson, *English binding before 1500.*
HOBSON, *Panels*	G. D. Hobson, *Blind-stamped Panels in the English Book-trade.*
HOBSON, *Abbey*	G. D. Hobson, *English Bindings in the Library of J. R. Abbey.*
S.S.L.B.	J. B. Oldham, *Shrewsbury School Library Bindings.*
WEALE	W. H. J. Weale, *Bookbindings and Rubbings of Bindings in the South Kensington Museum.*

LECTURE I

I MUST begin with two preliminary remarks. First, I propose to omit completely the whole subject of panel stamps, as they have been exhaustively treated already in Mr Hobson's monograph, and also the early Romanesque bindings which are a subject by themselves, and which have been fairly fully treated by him on several occasions; my remarks apply only to English blind-stamped bindings—stamps and rolls— of the fifteenth to seventeenth centuries. Secondly, several of those interested in the subject have asked me to start from the beginning, and not to be afraid to deal with quite elementary points. After much hesitation I decided to adopt this suggestion in case there are some who have in the course of their work to deal with bindings, but have been too preoccupied with other branches of bibliography to acquaint themselves with what has been written on this one.

The study of blind-stamped bindings should be based on rubbings combined with careful notes, and supplemented by a study of documents. I must admit at once the inadequacy of my knowledge of the last-named, as, living at Shrewsbury, it has been impossible to do very much more than make use of such documents as have been printed; it must be left to someone else to identify the craftsmen about whose work I may speak.

Rubbings, in which I find the best results can be obtained by the use of a fairly hard pencil (which does none of the harm that timorous librarians sometimes think) and thin but strong paper, are more valuable than photographs, as they show the exact size of tools used, and, given the luck of a good impression, they bring out detail, and make it recognisable, in a way that neither a photograph nor, curiously enough, the binding itself will do. But rubbings alone, of course, are not enough. Notes, not only of dates, but of inscriptions in the book or of fragments found in it, or of anything peculiar in the forwarding, are essential for providing additional internal evidence of the place and date of binding. Much time may seem to be wasted in carefully annotating the rubbings that one takes before putting the book away, but it is nothing to the exasperation engendered by finding a rubbing taken in less experienced days, which, because of lack of adequate notes, leaves one's investigations in the air. Then, when the study of rubbings and notes has made possible the grouping of bindings, the investigation of a College's accounts or of other such documents, may make probable, or at any rate possible, the attribution of the group to a particular binder.

The object of the study of bindings may be either artistic or historical, but the two viewpoints overlap, for any consideration of the artistic development of bindings, their designs and the tools used on them, and its relation to iconography in other crafts, necessarily involves first the dating and localising of them. Incidentally let us admit at once that it is only a small proportion of the mass of blind-stamped bindings which can claim much artistic merit.

Elementary though it may be, and already much discussed, the methods by which an attempt may be made to discover when and where a binding was executed are worth setting out in some detail.

I will begin with the problem of dating.

Normally, of course, one would say that the date of the binding was at any rate subsequent to that of the book which it covers, though by how much is a more difficult question. But there are two exceptions even to this. One is the case of an *emboîtage*, that is to say, a book that has been fitted into an old binding, or into part of one. To recognise this is merely a matter of using common sense, but it is quite possible at times to fail, by carelessness, to notice the likelihood of the binding being older than the book. The second is the case of some blank books, that is, books which consist of blank leaves used for ledgers, registers and other such purposes.

Now blank books are rather a snare, for they may be invaluable to us in dating a type of binding, but they are also capable of being very misleading. For the natural assumption would be that a blank book can be dated more exactly than any other, because one would expect that when a clerk responsible, for instance, for keeping accounts, was getting near the end of his book, he would order a new one, and that therefore the date of the binding on a blank book would usually be a little before that of the first entry in it. And in fact that generally was the case, as can be proved by the notes that some contain which give the date of the binding or of the purchase of the book. The date, for instance, of the purchase of the first account book of Shrewsbury School is ascertainable from an item in the accounts, while in the muniments of Winchester College there is a whole series of books containing notes of when they were bought or bound, which confirm this assumption and, incidentally, provide the key to the problem of the H.R. and I.R. family of binders.[1] But caution is needed, for sometimes what appears at first to be a blank book is not really one at all, and examination will show that the writing and ruling must have been done before it was bound. There are also confusing instances, such as a volume of Ordinances belonging to the Merchant Taylors Company of York. The writing in this dates from 1591, but the binding bears the W.G.-I.G. roll[2] which was never used after the first quarter of the century. The solution appears to be that the book belongs to the earlier period, as Sir Hilary Jenkinson, who examined it and told me originally of it, found signs of some earlier leaves having been removed and replaced by leaves on which is the present later writing. An even more extraordinary case is that of a blank book of Bruges quoted by Verheyden as having remained unused for fully 150 years.[3] A book in the Record Office,[4] bearing a roll not known to have been used later than 1563, but the writing in which begins in 1668, is probably another example of the same freakish circumstances.

When we turn to the bindings of printed books the usual and quite legitimate practice is to assume that, when an appreciable number with similar bindings bear dates in their imprints very near one another, those are the approximate dates of the bindings which were, it may be reasonably argued, put on soon after the books were published. If one goes through a College library one frequently comes across groups of books all bound alike, and printed in various places but at approximately the same date. This suggests that a member of the College has been sent to one of the book fairs to acquire all the latest books, and has brought them home and had them bound by the same local binder.

---

[1] See pp. 32–3. [2] No. 40.
[3] P. Verheyden, *Catalogue of Bookbindings Exhibition*, Antwerp, 1930, p. 196, no. 282.
[4] Record Office, E. 403/2433.

ENGLISH BLIND-STAMPED BINDINGS

For of course it has to be remembered, as Mr E. P. Goldschmidt[1] emphasised long ago, that normally fifteenth- and sixteenth-century books were imported in sheets and bound for the purchaser. It was the normal practice, but obviously there were a good many exceptions. Duff, for instance, remarked[2] on the fact that all the copies he knew in contemporary bindings of Henry VIII's *Assertio* were bound by Reynes (though actually there are some that were not). This could not have been a coincidence; clearly these were sold ready bound. Everyone, too, admits that stationers kept in stock for inspection a certain number of copies already bound in what are usually called 'trade bindings'. For some books, moreover, such as the Bible, different prices were specified by proclamation according to whether they were sold bound or unbound. And again, the well-known Act of 1534 not only forbids the importation for sale of books ready-bound, which it would not have done if the practice did not exist, but specifically states that 'great plenty of printed books...some bound in boards, some in leather and some in parchment' have been imported to be sold retail. And Christopher Barker (quoted by Duff)[3] asserted in 1582 that there had been in Henry VIII's time and 'to this day' stationers who 'buy their books in gross...to bind them up and sell them in their shops'. But no doubt, even before the Act, books more commonly awaited purchase before being bound, and this must account for the majority of the cases where one finds a book whose binding fits in with a group of a certain date, but bears a very much earlier imprint. And Mr Hobson proved,[4] with elaborate detail, that medieval manuscripts, too, often remained for long periods unbound.

Obviously no rule can be laid down as to how to determine, with a list before one of the dates of books similarly bound, what are the limits of the period of the binder's activity; one must always be prepared to think again when several books turn up with dates outside the limits that one had been inclined to assign. But the possibility has then to be considered that, in spite of the use of the same tools, they are the work of two successive binders, either because the span of dates, following in a reasonably close sequence, and thus suggesting contemporary binding, is too long for one man, or because some difference in style can be detected between the earlier and the later examples. This, of course, is a question which every student of bindings has constantly to face when examining a number of examples that are in some ways alike and in other ways different.

The rough-and-ready method of dating bindings from the dates of the books that they cover can sometimes be supplemented in other ways. A volume may contain an owner's name and date written either on the title or on the endpapers. If they are written on the title it does not help much, though it is arguable that this implies that at that date the book was in an unbound condition. But this is not necessarily true, as all book owners are not careful to confine their notes of ownership to the blank leaves. I could quote a relevant case where the first and last leaves are missing, and evidently were when the owner inscribed his name, for the dirt and discoloration on the first and last surviving pages on which it is written show that it was unbound at the time, and that the present binding was put on subsequently to the owner's inscription. The book,[5] an Erasmus in Shrewsbury School Library, is an interesting example of what can be learnt from such evidence. The owner of the tattered and then unbound volume was 'George James of Ratlinghope in the Countie of Sallope'. But the endpapers of the present

[1] Goldschmidt, pp. 36–40.   [2] Duff, *Westminster*, p. 119.
[3] Duff, *Westminster*, p. 76.   [4] Hobson, *Before 1500*, p. 56.
[5] Shrewsbury School, A. VI 20 (n.p. and d.).

binding consist of the sheets of an unrecorded pamphlet called *A Mirrour for Murtherers or a caveat for disobedient children*, which deals with a murder which took place at Bishop's Castle in Shropshire in 1633. On these endpapers there is a note in the handwriting of Chaloner, the Headmaster of Shrewsbury of the time: 'I bought this booke of Mr William Poulter the 18th of June 1636, price o. 6. 6. I houlde it worth at lest o. 12. o.' As the endpapers of a book are part of the binder's work it is possible to date this binding between 1633 and 1636. Further, Poulter is known to have been a Shrewsbury bookseller, and one of the rolls on the binding appears on another book[1] in the School Library which has written on the presentation label (which is dated 1625), but obviously added to it in a not much later hand: 'Bound by Mr Gittins.' This bald statement, with no addition to show who 'Mr Gittins' was or where he worked, suggests that he was a man who would be readily recognised by anyone who read the note, in fact, no doubt, a Shrewsbury binder. And though no binder of that name and period has been found in the documents, Gittins is a well-known local name. That he was a Shrewsbury binder is confirmed by the fact that, out of the nine known volumes bearing the rolls in question, seven are in the School Library, and in every case but one (which is probably the work of an earlier London binder) there is evidence that is consistent with the idea that the present bindings were put on at Shrewsbury at about this period. And no doubt the dishevelled book, belonging to George James who lived in a neighbouring village, was disposed of by him to Poulter in Shrewsbury, at whose shop Chaloner saw it, thought he had a bargain, bought it and had it bound by Gittins, who used up waste sheets of a pamphlet which would be more likely to be in Shrewsbury than anywhere else. While therefore inscriptions on the leaves of a book do not usually prove anything conclusive about the date of the binding, a date on the endpaper does prove that the book cannot have been bound after that date.

There are other considerations which help one to decide whether a binding is as early as the date of imprint. The use of a roll-produced decoration of course excludes a very early attribution. Paper, as opposed to wooden, boards are said by Mr Gibson[2] (and other people's experience would probably confirm it) not to have been used generally in England until the second quarter of the sixteenth century. Exceptions can be found, but one would certainly hesitate before attributing a binding, whatever the date of the book inside, to an early date if its boards were made up of paper. The same applies to paper pastedowns; I should not like to give a date for their first appearance, but the pastedowns of earlier books were almost invariably either fragments of manuscripts or plain vellum, which were certainly not, as Duff thought,[3] the mark of an Oxford binding; the use of vellum was far commoner in the fifteenth century than was thought when Hobson instanced its use as a characteristic of the Unicorn Binder; in fact I should say that in England at any rate it was the commonest form of pastedown at that period. A contemporary title on the back is, of course, a sure sign of a later binding.

Lastly, it is sometimes possible to check the date of a binding from the tools used, usually as a result of alterations by damage or re-cutting, at a date that can be approximately fixed.[4] This is more common with panel stamps than with rolls or small stamps, because panels were frequently re-cut, as for example the H.I. Pieta and Image of Pity panels which had a quite unaccountable 'A' slipped in above the lettering, the Egmont panel in which the 'S' before St Mark's name was changed into the binder's

[1] Shrewsbury School, G. VI 33 (Louvain, 1486).
[2] *Transactions of the Bibliographical Society*, vol. VIII, p. 29.
[3] Duff, *Westminster*, p. 104.　　　　　[4] See Pl. I.

4

initial 'E', Reynes' double panel in which a cock was substituted for a pomegranate, Reynes' Baptism panel in which tufts of grass and milling in the border were added, and the numerous panels which appear sometimes with, and sometimes without, 'nail-marks', if this is what the curious round depressions at top and bottom of the panels really are. An example of accidental damage to a panel is the cracked one with heads in medallions from Louvain, the damage to which Husung dated as 1535.[1] Cases of deliberate alteration to rolls are those where the binder's initials have been obliterated by a subsequent owner, the best known of which are Reynes' roll whose I.R. was erased after his death in 1544, and the roll signed L.V.L. which, with the initials obliterated some time after 1509, is Siberch's roll numbered III by Gray. Similar are the cases where Spierinck changed the N.G. on his own roll, and the I.S. on Siberch's roll to N.S. Where the date of these changes can be fixed, the knowledge may be valuable in dating or attributing a binding. An example will best explain this. The lozenge dragon stamp of the Dragon Binder sometimes has, and sometimes has not, a nick under its tail. All the books except one printed in or before 1504 show the tool in its first, un-damaged, state, those of 1505 and 1506 in its second state. It is reasonable to suppose that the damage was done in 1504 or 1505. Now there are seven volumes[2] which bear a roll in addition to the dragon stamp, and as this is not in accordance with the Dragon Binder's usual practice, it might be thought that they were the work of a successor. But on five of them the stamp is in its first state, and on account of certain peculiarities of style it is beyond doubt that the tool was still in use by the Dragon Binder after it got damaged; these books with rolls on them must therefore be his work.

There is one other example that may be cited, but it is an unusual one, for it is a case of dating bindings not from the state of a tool when used, but from the use of one tool rather than another. Reynes used a well-known signed roll (no. 432), with lilies, falcon, bee and dog, and with it a pineapple stamp (no. 437). This stamp appears habitually also with a less imposing unsigned roll bearing similar decoration (no. 433). But there are numerous rolls with almost exactly—in one case (no. 510) exactly—the same assortment of flora and fauna, and it was only on closely examining them that I realised that these two were in every single detail the same, only one was the reverse of the other. I then noticed that, of all the vast number of examples of the unsigned roll, not one was on a book printed after 1520, whereas all but very few of the examples of the signed roll were printed after that date. It seems quite clear that Reynes not merely gave up abruptly using the unsigned roll in that year, but actually lost it, and sent one of his bound books to his die-cutter with instructions to copy the roll, but to make it larger and add his initials, with the result that the die-cutter, copying an impression on leather, or a rubbing of it, produced a design in exact reverse of his model. Incidentally, this is the more likely explanation of tools which are the reverse of one another, and even of some of those with lettering in looking-glass writing, and not the explanation usually given, that they must be the work of the same die-cutter; in fact they are probably generally by someone else. What I have stated here seems to me a perfectly sound reason for dating any binding bearing Reynes' signed roll not earlier than 1520.

When we turn from the dating to the localising of bindings, there are more signposts to point the way to a right conclusion, but even so, to lay down dogmatic rules would be unscientific, and we must be prepared to modify or even reverse them if we run up

[1] M. J. Husung, *Otto Glaunung Festgabe*, 1936.
[2] Merton College, Oxford, 53. hh. 3–4 (Venice, 1501) and 57. hh. 5–8 (n.p., 1501), and Magdalen College, Oxford, F. 13. 10 (Venice, 1495).

against exceptions. For instance, I should agree that most well designed and executed panels were not English products, but to go further and say that because a panel is well designed and executed it cannot be English seems to me unscientific and arguing in a circle. If in fact it is English, it simply means that, while the quality of design and execution may have been legitimately taken into consideration in assigning its origin, we have to accept the fact that that consideration is not conclusive evidence, and our rule must be taken as being of less universal application while still of some validity. This may seem obvious, but it points to what I believe to be a misleading method of reasoning that has sometimes been employed.

In trying to localise a binding which does not provide us with a self-evident clue, we may start with the place named in the imprint, or the place where, if it is a manuscript, the book can be proved to have been written. This will not usually help us much, owing to the fact already remarked on, that books—even manuscripts—were sometimes left long unbound, and were usually bound to the order of the purchaser. But if the book is, and is known to have been from an early date, in England, and was printed in this country, the chance of its having strayed abroad (especially in view of the apparent lack of two-way traffic in books between England and the Continent), been bound abroad, and found its way back to this country at an early date, is too remote to be worth much consideration. The same applies to a limited extent to towns when the place of printing or writing was also a known centre of binding, such as London, Oxford or Cambridge, though one must always allow for the possibility that the book was sent to the metropolis to be bound, of which a number of Canterbury Registers, bound in London, is an example. A slightly different point is the argument that may be drawn from a considerable number of examples of one binder being found in England, especially if they are widely distributed. Though plenty of foreign-bound books of course exist in this country, if books by one binder exist here in any large number, except where they form a group obviously imported by one owner, like the Schevez books[1] at St Andrews, it is reasonable to think that the work was probably done in England. On the same analogy it is legitimate to notice in what old libraries the books now are; one naturally expects to find many Cambridge bindings at Cambridge, and many Oxford bindings at Oxford, and, with caution, one can apply the argument the other way round.

A rather similar argument may be drawn, as Mr Goldschmidt has pointed out, from early notes of ownership or of currency written on the endpapers which prove in what country (or perhaps town) the book was soon after it was bound. As in the case of dating a binding, an inscription on a printed page is of less importance. A book in England with French notes on the title was pretty certainly once in France, and may have been bound before or after it came to England, but a book with French notes on the endpapers was equally certainly bound in France. And the importance of printer's waste found either making up the boards, or forming the endpapers, or used as a reinforcing piece, was first pointed out by Weale[2] as being very strong evidence of the place of binding, and Mr Goldschmidt has further elaborated the argument.[3] But both pointed out that while printer's waste, that is, trial sheets printed on one side, or other ephemera of a printer's office, is of much importance, because it would be likely to be found near the home of the printer, bookseller's waste, for instance old books or fragments pulled to pieces as scrap paper, might have come from anywhere, and conclusions drawn from its use might be very misleading. A similar importance attaches to pieces of deeds which are very frequently found used as reinforcing pieces, and, not only

---

[1] See Hobson, *Cambridge*, p. 36.　　[2] Weale, p. xix.　　[3] Goldschmidt, pp. 119–21.

from their language or the mention of a sovereign, but owing to the fact that they usually contain many names of persons and places, they may be of the greatest evidential value.

And here perhaps I may be allowed to digress for a moment to put in a plea for the most conservative possible treatment when books are repaired. I would not myself go so far as to say that a book falling to pieces should never be repaired; that seems to me a counsel of perfection. But if it is repaired, even if the sewing cannot be retained, absolutely everything else ought to be. Fragments of MS. or printing which may provide the student with evidence must never (save in the most exceptional circumstances) be separated from the book. If the MS. experts want to read both sides of a MS. pastedown, it can be raised and left as a fly-leaf. If the leather of the binding will not stand up to use, every scrap of it, including the back (a point the importance of which is often missed) must not be neatly trimmed at the edge, but replaced exactly where it was before on the new binding, and except where beetle has made it impossible, the old boards should certainly be retained, and of course without any new endpapers with which so many repairers love to tidy up an old book. This warning is necessary and needs to be emphatic, because, even in some libraries of distinction, history is being sacrificed to neatness or even convenience.[1]

Identity of rubrication has sometimes been thought to be conclusive of the identity of the binder of two or more books, on the ground that binding and rubrication were done at the same time. This seems to me to be going much too far, because, even if the binder did sometimes do the rubrication himself, it is unlikely to have been common, as the techniques of the two crafts appear to have no relation one to the other. Moreover (to quote Mr Hobson)[2] 'it has been conclusively shown by Dr Husung' that in many cases manuscripts were bound before they were rubricated, and Duff has stated[3] that in nearly all the law books printed by Lettou and Machlinia the initials appear to have been the work of the same person, though if they had been the work of the same binder he would certainly have remarked on it.

Passing from considerations of this kind, which are largely a matter of enlightened common sense, we must come to arguments which deal more with the technique of sewing, forwarding and finishing. This is a matter of the teaching of experience, and here the warning against dogmatism needs the greatest emphasis. For the conclusions that I reach, however cautiously expressed, may well be invalidated by someone else's wider experience. And I ought to add that I have not studied bindings anywhere outside libraries in Great Britain.

The sewing of a binding, unfortunately, does not help us much, for the simple reason that little can be seen of it unless the book has come to pieces. But in any case, I am told by Mr A. Birdsall, a binder of wide experience in old books, that, except for the quality and colour of the thread used, there seems to be little difference between the sewing methods of English and foreign binders of the earlier period. I am not, of course, thinking of very early books, such as the Coptic bindings, which had peculiar features of their own, which I have recently found persisted almost unchanged to modern times in the district from which they came.

But when we come to forwarding there are many features which must be noticed when one is trying to assemble arguments for deciding where a book was bound. In England the most common wood used for boards was oak, in Germany beech; in

---

[1] Mr Roger Powell has suggested to me that, when a book has been repaired, much of the history of the binding can be preserved if the repairer inserts in the book an exact statement of the work that he has done on it.

[2] *The Library*, 1934, p. 170.   [3] Duff, *Westminster*, p. 51.

Scotland pine was sometimes used. There is at Shrewsbury an early seventeenth-century English binding[1] which is very remarkable because it consists of beech two-ply, and I have never heard of another example. The thin boards used, about 1 mm. thick, are not placed, as in modern two-ply, with the grain at right angles, but with the grain running the same way and, for some reason, a piece of paper inserted between them. The leather used in England for blind-stamped bindings was usually calf, and only rarely sheep, whereas both were used in France and the Netherlands; pig and calf were the most common materials used in Germany, and morocco in Italy and Spain. A reddish brown leather usually means a binding of the Lower Rhine. Mr Gibson has mentioned[2] that many of the bindings in the fifteenth-century library of the Duc de Berri are recorded in the inventory as of 'cuir vermeil'. This may mean what it says, but it is interesting to note that quite a number of the books in Shrewsbury School Library, still in their original stamped bindings of ordinary coloured calf, are recorded in the seventeenth-century list of accessions as 'covered with red leather'; either the meaning of the word 'red' has changed, or the cataloguer was colour-blind. A pink pigment is a different thing from a leather tanned red; the former is essentially a Cambridge practice, though there is at Lambeth one binding[3] which, except for this, I should unhesitatingly attribute to London; probably it is a freak.

A much more important piece of evidence comes from the position of the clasps. There are exceptions, but they form a minute percentage in my experience to the rule that in England and France the clasps were put on the upper cover and the catches on the lower, and that in Germany and the Netherlands the clasps were on the lower and the catches on the upper. Hebrew books were naturally treated the other way round because the 'lower' cover was next to the first page. After allowing for books that have been rebacked, and the boards inadvertently changed round, I believe that the exceptions to this rule are so few that, though it leaves open a considerable area of doubt between English and French, and between German and Netherlandish bindings (Italian and Spanish are pretty easy to distinguish otherwise), it is one of the most valuable clues to nationality. But there were not always clasps. There were, however, often silk ties, and if these appear at head and tail as well as fore-edge, the binding is nearly always French or Italian, though I do know two instances[4] of this use in England. Sometimes Italian and Spanish binders used clasps at head and tail instead of ties; as far as I know no one else did. I have not found Duff's rule[5] about the number of bands a safe guide for nationality, but French and Italian binders tended to use more bands than others. The use of leather headbands 'stabbed' through the leather at the back is an ordinary early English practice, and plaited leather headbands, usually pink, are more common on the Continent. An English binder who was exceptional and used plaited headbands was the Greyhound Binder.[6]

On English bindings a slight bevel all round the boards was common, while a sharper one was characteristic of the Netherlands. Mr Gibson says[7] that many Italian bindings had a bevel on the inside, and certainly this was often so in England; a craftsman[8] who worked for Eton early in the seventeenth century had an unusual habit of running a roll along the inside bevel, as did also sometimes Van Hagen who worked at Aberdeen in

[1] Shrewsbury School, D. VIII 1 (Antwerp, 1603).
[2] *Transactions of the Bibliographical Society*, vol. VIII, p. 28.
[3] Lambeth Palace, 87. c. 16 (Rouen, n.d.) bearing Royal Arms panel (Weale, R. 171).
[4] Brasenose College, Oxford, Registrum A (illustrated, Gibson, Pl. XXI) and Bodley MS. 523.
[5] Duff, *Westminster*, p. 122. [6] See p. 21.
[7] *Transactions of the Bibliographical Society*, vol. VIII, p. 29. [8] Using roll HM. *b* (2).

the 1630's. But I think a bevel on only the middle portion of each edge is always the sign of German work,[1] though whether the practice was confined to one part of Germany I am not experienced enough to know. A doeskin covering of the back fastened down by strips of brass was used at times in England and Germany and perhaps the Netherlands, but not to my knowledge in France. Abroad, metal bosses were common, especially on the lower cover. They were sometimes used in England, but they were not usually of the flat-topped variety;[2] from the beginning of the seventeenth century in fact they generally form part of a decorated square metal cornerpiece (one I know of bearing the Tudor emblems on it),[3] which differs from the German type of cornerpiece, the corners of which usually project towards the centre of the book at an acute angle. I have myself never succeeded in disentangling the different types of endpapers—vellum, paper, manuscript and printed fragments—but sufficient assiduity applied to the study of these and other things, such as the colour of the edges, the appearance of cord marks on the back, the designs of the clasps (complicated by the fact, proved by customs rolls, that they were sometimes imported), and the watermarks on the endpapers (which necessitates a knowledge of the paper trade), might well bring in these, too, as evidence of nationality or district.

Lastly, there has to be considered the guidance to be obtained from the finishing of the binding, probably the first thing that most people quite properly look at. Finishing falls under two heads, the design, that is, the way in which stamps and rolls are arranged, and the actual tools which are used.

In regard to design, as is well known, certain countries and some towns have one or more patterns which are more commonly used than others, and certain of them are comparatively rarely found elsewhere. The three types of design most commonly used in England, and not exclusively in any particular town, are:

(a) In the fifteenth century, an intersecting frame of fillets, containing stamps which are more often not carried beyond the points of intersection (to do so seems to be almost a peculiarity of Cambridge), the space within being divided by diagonal fillets into lozenge and triangular compartments, each containing a stamp; the sixteenth-century version is the same but with a roll-produced frame that is intersecting.[4] This pattern appears also in the Netherlands, occasionally in France, and (usually in a modified form) in Germany, but I think the intersecting type of frame, whether of stamps or of rolls, is rarely found on the Continent.

(b) In the fifteenth century, a frame of stamps, not usually intersecting, with the whole space within filled by concentric frames or vertical (or, at Oxford, horizontal) rows of stamps;[5] the sixteenth-century version of this is an intersecting roll-produced frame, filled entirely by vertical strips of rolls—a pattern much favoured at Cambridge.[6]

(c) In the sixteenth century, a frame, not intersecting, formed by one or more rolls, with either a plain space within, or a panel separated from the frame by a plain space, consisting of vertical strips of rolls,[7] more often in England not surrounded, as it usually is in France, by an inner frame.

But there are many other designs which occur on English bindings quite often, for instance an outer and an inner frame with a space between them,[8] which is a design indistinguishable from one often used in France. When, however, the space inside the inner frame is divided into three by horizontal fillets, or a pair of horizontal fillets join each of the bands to the outer frame,[9] I think that, if English and not German, it is

---

[1] See Pl. VIII.   [2] See Pl. VII.   [3] Bristol Central Library, EPB. 430 (n.p. and d.).
[4] Pl. II, 1.   [5] Pl. II, 2.   [6] Pl. II, 3.   [7] Pl. II, 4.   [8] Pl. II, 5.   [9] Pl. II, 6.

always a London binding.[1] A roll-produced lozenge panel inside the frame,[2] especially if accompanied by horizontal lines from the bands, is nearly always a sign of London work, but the lozenge was occasionally used at Oxford, though I think never at Cambridge. It is occasionally found in Germany and Spain, but the lozenge is then generally squarer. A common Oxford design has the space inside the frame divided into four triangles by intersecting rolls set diagonally in the form of a saltire;[3] I know one use of this design by Reynes, three by Godfrey and one by Spierinck, and two or three others which appear not to be Oxford work, but they are rare. Quite exceptional English designs are: rolls being used instead of fillets—I know some half-dozen binders who do this[4]—to divide the panel in the London fashion into three (or, by the binder I.R.[5] into as many as six) sections; another, the inner panel, whether rectangular or lozenge, being mitred by a roll to the corners of the outer frame;[6] and another, very rarely used, the space within the frame being divided into two or more panels by the vertical use of a roll. All these last have been cases where rolls have been used, but in the fifteenth century, with the greater freedom of design given by the use of stamps, all sorts of other patterns were invented. Some English binders[7] who had a book in several volumes to bind, showed a touch of originality in using a different design on each. Different designs on the two covers were very rare in the sixteenth century, but Mr Hobson certainly overstated it when he gave as a distinction between Romanesque and later binders the fact that the former used a different design on each cover and the latter generally did not; it is true of the sixteenth century, but in the fifteenth different covers are by no means uncommon.

Some bindings with these designs may obviously be confused with foreign productions. The typical Paris over-all design,[8] with a border right against the edge and no plain spaces left, is easy to recognise; I know only two English binders[9] who used it, and they not often. The panel-stamp binding with large stamps set round the outside of the panel is more often French; in England the panel is generally found alone, or rarely surrounded by a roll. Netherlandish fifteenth-century bindings with a frame of stamps enclosing lozenge and triangular compartments are often difficult to distinguish from English, but when the frame is composed of alternate round and lozenge stamps[10] it is more likely to be Netherlandish, as also are bindings in which the frame is made up of a number of what I have called crocketed cresting stamps, with a rose in each corner,[11] a pattern never, I think, used outside the Netherlands and Lower Rhine district. Again, English binders never used,[12] as far as I know, the large wooden matrices, instead of metal, for panel stamps, which can usually be recognised by the broader treatment of the cutting;[13] these Mr Verheyden said were peculiar to blank books from Bruges and Ghent; they are not very common, and I know only three examples[14] that have strayed into England.

---

[1] There is one example of this design which may be Norwich work. [2] Pl. II, 7. [3] Pl. II, 8.

[4] The binders I.R., F.D. and F.P.-W.L., the Garland Binder (*S.S.L.B.* p. 18), and the binders who used rolls HM. *b* (2), HM. *h* (11), FL. *a* (2), DI. *d* (2) and HE. *b* (6).

[5] See Pl. XXX. [6] Pl. II, 9.

[7] E.g. the binder F.D. and the binder who used roll RC. *d* (2). [8] See Pl. III.

[9] The binder who used roll RP. *a* (1) and the binder who used rolls FP. *g* (2) and RC. *b* (6).

[10] See Pl. IV. [11] See Pl. V.

[12] Unless the remarkable St George and Tudor emblems panels on Bodley MS. 523 (reproduced in Brassington, *Historic Bindings in the Bodleian*, Pl. VII) are examples of this technique.

[13] See Pl. VI.

[14] Longleat MS. 266, account book of the Drapers' Company, Shrewsbury, and a single cover in the Victoria and Albert Museum.

German fifteenth-century bindings are often similar to the English ones with either vertical rows of stamps (especially in the Cologne district), or a frame with diagonal fillets, or more commonly with a characteristic curved stamp,[1] to form the lozenge compartments. But they can usually be distinguished (apart from the frequent use of pigskin) by, in the former case, a greater number of small stamps being used, and in the latter by the use of large and rather heavy foliage stamps. In the sixteenth century when panels are used, there is almost always in Germany a plain space, occasionally containing two or three foliage stamps, or more commonly the owner's initials and the date, at top and bottom of the panel but not at the sides.[2] Italian and Spanish blind bindings are so unlike English as to be easily distinguishable, with the former's characteristic leafy or rope-work frame, and the latter's rather unusual and fantastic stamps, always in intaglio, scattered about in a less methodical way than elsewhere.

A few small details may be mentioned. Hatching on the half-bands or above and below them is characteristic of Oxford, but was also often used in France. Hatching on the edges at head and tail near the back was used at Oxford and Cambridge, but probably in this period nowhere else. Fillets down the back are usually a Cambridge feature, but are not so peculiar to that centre as used to be thought.

It need hardly be said that this account does not profess to exhaust all the local characteristics of design; nor must these tests be taken as infallible guides; but, taken with other evidence, they may help cumulatively towards a conclusion, though when one has convinced oneself that some feature is a characteristic of a particular place or a particular binder, too often one finds that binders are distressingly apt not to run true to form.

It remains to consider briefly what can be deduced from the actual tools used, and here even greater caution is needed, for two reasons. In the first place it may be difficult to identify with certainty tools that are close variants of one another, because of the varying state of the tool when used at different times, because of the varying condition of the bindings, and because leather moistened to receive an impression shrinks when it dries to a varying extent, so that impressions of the same tool vary in size—it is said by as much as 20 per cent., though I have never myself seen a variation anything like as large as this. When there is doubt, stamps can best be compared by taking rubbings of them. But unless the identity of a tool used on two bindings can be established beyond doubt, the risk of building theories on unsound foundations is so great that it is better to ignore completely the presence of that particular stamp in drawing any conclusions.

Secondly, in assigning a binding to the craftsman who produced another bearing the same stamps or rolls, it has to be remembered to what an extraordinary extent tools passed from one hand to another. It is quite useless to put together a group of stamps and then be satisfied that all bindings bearing any of them are the work of the same man. I know such a group, consisting of stamps, each of which is used with at least one other of the group, which numbers 139 tools. Nothing approaching this number of tools was ever used by a fifteenth- or sixteenth-century binder. On the death or retirement of a binder his tools were bequeathed or dispersed by sale; in the latter case it is reasonable to suppose that they were more likely to be bought in his own town than elsewhere, which is sometimes the basis of an attribution of certain binders to particular towns; but there were clearly some exceptions. And the binder himself might move, and the tools become associated with another town. Notoriously this was the case with foreign

---

[1] See Pl. VII. [2] See Pl. VIII.

binders bringing their tools to England; it is very common with panels, less so with stamps, and least with rolls. The reason for this may be the greater value of the panels, and the fact that most of the immigration had already taken place before rolls generally superseded the use of small stamps.

But when all necessary allowance has been made, it is legitimate to regard some types of tools as associated with certain countries or centres.[1] For instance, I think that a lattice stamp was never used outside England, and almost the same may be said of the bird, wivern and lily type of roll (though I know one indubitable case of its use in France), which is odd, as it is exactly this design that forms the border of so many French panels. Very many almost exactly similar rolls were used in different countries (of which I shall be speaking in another lecture),[2] but in some instances there is some small feature which seems to distinguish those used in one country from those used in another. Rolls which are to be looked at not from the end but from the side (which I call 'side-view rolls') are used rarely except in England and France, but when they are divided up into small panels they are nearly always French, and when they are continuous and unbroken they are nearly always English, though I know one,[3] which I shall speak of later, that is German, and one used in England[4] may have come from the Netherlands. What are usually called Renaissance ornament rolls, used in England and France are, for obvious reasons to be mentioned later, extremely similar, but if they contain elongated dragons addorsed I think they are always French. French, too, more commonly are cherubs with their necks divided into two and prolonged to a considerable extent; the elongation of the English cherub's divided neck is usually less pronounced. Rolls and stamps which, on the leather, appear in intaglio, though by no means unknown on English blind-stamped bindings (of course all gold stamps are in intaglio), are far commoner in Germany and also in Scotland. And as between English and French similar rolls bearing Renaissance ornament, conventional foliage or twisted stems with pineapples, on the whole the broader ones are likely to be French.

About more exact localisation there is not much to be said. The chequered type of stamp is almost peculiar to Canterbury; there is a type of lily stamp that seems specially associated with Cambridge, and groups of small circles are considered to be peculiar to Oxford. The use of the briquet Verheyden regarded as conclusive of a Ghent or Bruges binding, and no doubt it generally is, but it was occasionally used in England, I think at Oxford.

But very many tools of the same type, and curiously enough not only the obvious sort of commonplace ones, are used in different countries and show no special local characteristic. And when all these varying kinds of tests that I have suggested have been applied in an attempt to localise a binding, one has too often to give up the problem as insoluble, and confess oneself beaten.

[1] See Pl. IX. [2] See Pl. XXXIII and p. 35. [3] No. 521. [4] No. 924.

LECTURE II

BEFORE turning from ways in which help may be gained towards dating and localising a binding to the consideration of some of the individual binderies, one or two preliminary observations may be made. One relates to the number of binders there may have been working in England in the earliest part of the period under review. Mr Hobson, in his Sandars Lectures, estimated the probable number who were working between 1450 and 1509 as from thirty to sixty, and he named eighteen specific binderies of that period as definitely English, and a further fifteen as probably or possibly English.[1] I feel sure that this estimate is far too conservative, and it is only natural that in the twenty odd years since it was made, many other binderies should have come to light. Including Mr Hobson's eighteen and most of his 'probables', I could name the work of some sixty distinguishable and definitely English binderies, stopping at the more convenient earlier date of 1500. In addition there are what I might call the 'throw-outs', that is to say, bindings on which tools are used which at one time belonged to a distinguishable bindery, but which were clearly not its productions. Such groups of 'throw-outs' I have not been able to attribute to any one bindery, but each involves the addition of one or probably more binderies to the list. Further there are several others that I should suspect of being English, without any sufficiently definite evidence, and to these of course must be added no doubt many other binderies of which no example has yet been found, or which have not yet been identified as being English. In view of the likelihood of its being soon put out of date, I will not myself risk an estimate.

Broadly speaking, and excluding panel stamps, blind-stamped bindings fall into two groups, the earlier ones that are decorated with small stamps, and the later on which rolls are used. The question is—for it may affect our judgement in dating a binding—when did the change take place? Of course the transition was gradual, and the stamp as a major form of decoration (as opposed to the use of insignificant stamps to supplement a design worked out in rolls) survived in exceptional cases long after the use of the roll had become general. Spierinck, for instance, used a stamp-produced frame on a book as late as 1521.[2] But the more important point to discover is how early may we allow ourselves to assume that an English binding bearing a roll is contemporary with the date of the book it covers. Obviously it is impossible to be precise. Here and there something like proof may be found of the fairly early use of a roll as, for instance, in the case already mentioned of a roll used on two books in five volumes dated 1495 and 1501, with the dragon stamp in its first state, that is, apparently before 1505. I have elsewhere[3] committed myself to the view that though this puts the use of rolls as early as 1505, the books were not bound as early as the dates within them. But I am beginning to think that this is not necessarily the case, and that the use of rolls in this country began

---

[1] Hobson, *Before 1500*, pp. 53–5. [2] Gray, Pl. XIV. [3] *S.S.L.B.*, p. 24.

13

earlier than I, for one (and I think others), was inclined to believe. It is possible airily to dismiss the view I now put forward tentatively, by saying that all the roll-produced bindings on earlier books were executed a number of years after they were printed, and such a statement no one is likely, except by some lucky chance, ever to be able to disprove. But one must take account of such facts as these: apart from even earlier examples, Spierinck's rolls are known on books dated in every year from 1499 except two, Godfrey's in every year from 1501 except one, L.V.L.'s in every year from 1500 except two, while Reynes' roll appears on several books every year from 1493, including twenty-three on books printed before 1500. At least three of these binders were aliens, but even before denization they were at liberty at that time to practise their craft; the unfortunate thing is that we do not know in any one case when the binder settled in England, though in Godfrey's case it was certainly as early as 1503. All things considered, it seems to me that rolls were probably in occasional use in England at the turn of the century, and possibly a little before, and that it is not necessary to regard as later work bindings bearing rolls on books printed as early as this. The latest date of a binding feature is easier to fix than the earliest, because comparatively few books were of later date than their present binding, and though rolls, especially those used at Oxford and those of the heraldic crested type,[1] strayed well into the seventeenth century, they probably ceased to be used blind, except of course the very narrow featureless ones which went on indefinitely, before 1660. I do not myself know an example later than 1654.

Of the large number of fifteenth-century binders, and of the immensely greater number that there naturally were in the sixteenth, it is only possible to comment on a few of those either that have not been dealt with by other writers, or who have already been the subject of study but in regard to whom a little more information is now available. It is of course necessary to scrutinise peculiarities of technique and style as well as the tools used before attempting to distinguish a particular bindery.

As in private duty bound, I will begin with Cambridge. The best-known peculiarity of Cambridge work is the use by several binders of a pink pigment, which, to judge by a number of extant examples, was probably smeared over the whole cover, though in some instances it survives now only in the impressed parts of the leather. I know only one case of its use apparently outside Cambridge.[2] I cannot believe from the appearance of these many books that Duff was right in thinking the colour was due to tanning; it seems clearly to have been put on after the book was bound. The complete filling of the space inside a roll-produced intersecting frame with strips of rolls[3] was also a common Cambridge feature, but I know one example of the practice on a Reynes binding, and there is one other London binder[4] who frequently used it. I do not think the design, common at Oxford, of concentric frames made up of stamps was ever used at Cambridge; stamps, when used, were generally arranged in the typical English way to form a single frame, and to fill the compartments within it that were formed by intersecting diagonal fillets. But it is noticeable that, unlike most contemporaries, three Cambridge binders, the Unicorn Binder, the Lattice Binder and W.G., all made their stamp-produced frames intersect.

Gray did valuable pioneer work[5] on Godfrey, Spierinck and Siberch, and as far as I know there is not much (except very many more examples than he quoted) to add to what he wrote. Godfrey's characteristic lattice stamp[6]—easy to recognise because of

---

[1] See Pl. LIV.   [2] See p. 8.   [3] Pl. II, 3.   [4] The binder who used roll SW. *b* (6).
[5] G. J. Gray, *The earlier Cambridge stationers and bookbinders*.   [6] No. 994.

its thirteen perforations, whereas most had nine, sixteen or twenty—was used by another binder before him whom I have called the Lattice Binder;[1] but his work is easy to distinguish, among other things, by the fact that Godfrey does not seem ever to have used pink pigment as his predecessor generally did. But as one stamp,[2] not given by Gray, a fringed foliage ornament, is used once with Godfrey's signed roll, it was probably once in his possession, though not for long, as it passed through the hands of several binders, including H.C. and possibly Siberch. In addition it is now known that the two panels signed G.G. belonged to Godfrey. Two of his rolls, one signed (Gray, III and V), passed after his death into the hands of a London binder. I know three bindings which I feel sure were Godfrey's work though they have the Oxford design of a doubled roll used in the form of a saltire, and one, at Deene Park,[3] where there is no frame, but the signed roll is used in the strangest manner, horizontally across the whole cover. This definitely is not, as might be imagined from the description, an *emboîtage*.

To the fifty-three bindings of Spierinck's mentioned by Gray, I have been able to add another 254, and these do add a little to our knowledge. For, on the ground that they had never been found with a roll signed N.S., Gray only tentatively attributed to him roll V, signed N.G.[4] and roll VI.[5] But as they were used together, and nine cases of roll V have turned up on bindings bearing Spierinck's signed roll, they may both be definitely put down to him. Moreover, these additional examples have introduced us to a signed roll not recorded by Gray, a diaper, like his roll II, but with his mark and initials introduced into every fourteenth lozenge and triangle.[6] There is also one example of his using the Heavy Binder's fleur-de-lis.[7] Mr Hobson was able to add two more panel stamps to the two commoner ones illustrated by Gray. There are a few more examples now known of Spierinck using Siberch's roll[8] with I.S. altered to N.S., and two of his using it unaltered. There are also two examples of Siberch's signed roll and Spierinck's floral roll[9] being used together, and two, besides the one mentioned by Gray, of Spierinck's and Godfrey's signed rolls[10] appearing on the same binding—which is all very confusing. I know one example of Spierinck using the saltire design, and there are a few cases of his using a three-sided frame, lacking the side next to the back. He fairly often used pink pigment all over his bindings. It should be remembered that though Spierinck did not die till 1545, Gray was no doubt right in thinking that he ceased binding about 1533, for there is a number of bindings bearing his floral roll which are clearly not his work, but apparently the work of a London binder, who, as Mr D. Paige pointed out to me, had the odd habit of running a single fillet from head to tail of the cover near the back.[11]

In spite of another thirty-six bindings bearing Siberch's rolls being now added to the six given by Gray, he remains, as a binder, as shadowy a figure as he did when Gray wrote. Gray gives him three rolls, one signed, one of dancing peasants, and one with animals.[12] Of the signed and unaltered roll, I know sixteen examples, including two on bindings bearing Spierinck's roll III and probably the work of the latter. The dancing peasants roll appears only once with Siberch's signed roll, five times alone, three times with Godfrey's fringed foliage ornament,[13] seven times with one or another of Reynes' stamps and twice with a fleur-de-lis of unusual form.[14] The animals roll, which

[1] See p. 18.   [2] No. 1006.   [3] Deene Park, VII *b* 12 (Basle, 1521).   [4] No. 748.
[5] No. 594.   [6] No. 27.   [7] No. 35.   [8] No. 750.
[9] No. 704.   [10] Nos. 561, 562.   [11] E.g. Caius College, Cambridge, K. 20. 15 (Basle, 1541), etc.
[12] Nos. 750, 924, 565.   [13] No. 1006.   [14] No. 1054.

turns out on examination only to be L.V.L.'s signed roll with the initials obliterated, is never used with the signed roll, but one book bearing it contains fragments of Siberch's printing and a letter addressed to him about 1523. The only other examples of this roll are seven when it was used with the fleur-de-lis stamp already mentioned or a merry-thought stamp,[1] or both, and one where it is used with the roll signed Z.C. and a floral roll.[2] Truly a tangled problem! Each of the unsigned rolls is thus connected with Siberch by one book only, and each is connected with at least one other binder. Even the signed roll, which is known to have passed into Spierinck's hands and been altered by him, was apparently used by him unaltered twice, and for all we know, it may have been more times. One can only conclude that it is necessary to be very chary in assuming that bindings, merely because they bear one of his rolls, were Siberch's work, especially as his career as a stationer was extremely short.

This exhausts the number of Cambridge binders of whom we definitely know both their names and their productions. But there are others whose bindings we can identify without knowing their names. Two who signed their bindings have been referred to already, L.V.L. and Z.C. (if I have read his initials aright). Of Z.C. I know only four examples, dated 1519–22, all of them, as it happens, in Cambridge, three being in College Libraries. This, and the fact that his roll once appears with Siberch's animals roll rather suggests Cambridge as his home, but there is no more definite evidence.

L.V.L.'s work—the books are dated 1500–9—is more intriguing, because his roll is associated with a number of other tools, some of them signed by other binders. These are W.G.'s signed stamp[3] (1507), W.G.-I.G.'s roll[4] (1503, but bound much later), and H.I.'s Pieta panel[5] in its first state (empty covers and therefore undatable). The only other stamps used with this roll are two pineapples[6] used later in London, a lattice stamp[7] and three otherwise unknown stamps,[8] a rose, a fleur-de-lis in a lozenge, and a very unusual type of rectangular stamp bearing a lion and wivern. None of the last three helps us at all. Nor can it really be said that the others help us, for they only make the movements of this roll additionally confusing. For it appears to have been used first at Cambridge by W.G., then by W.G.-I.G. and by H.I., both apparently Londoners, then again about 1523, with the initials erased, at Cambridge by Siberch, if we are to accept the strong evidence of the Siberch fragments already mentioned. All this seems very improbable, and with facts as at present known, the problem seems completely insoluble; but I think it is safe to call L.V.L. a Cambridge man.

As it has been necessary to refer to W.G.,[9] he had better be considered next. We may start with certain facts; not only are the W.G. stamp and the W.G.-I.G. roll used together at any rate twice, but the monogram on them is identical, and this cannot be a coincidence; they must both have belonged to the same firm or the same family. The dates of the books extend from 1478 to 1533, the single monogram being on the earlier books and the double on the later, which suggests that two generations are involved, a father and two sons, the elder of the latter probably bearing his father's Christian name and consequently using the same monogram. Two breaks occur in the sequence of dates of books at present known bearing these bindings (which number 121),

[1] No. 1027. [2] Nos. 879, 730. [3] No. 20. [4] No. 40.
[5] No. 9. [6] Nos. 960, 973. [7] No. 999. [8] Nos. 1035, 1051, 1074.
[9] See Pl. X. The following examples of this bindery are additional to those already recorded in Hobson, *Cambridge*, pp. 46–7 and *S.S.L.B.* pp. 58–9: (i) 1500 Venice: Deene Park, X *a* 8. (ii) 1513 Venice: St John's College, Oxford, Ξ 2. 20. (iii) 1513 Paris: Cambridge University Library, Rel. bb. 51. 1. (iv) 1518 Paris: Trinity College, Oxford, I. 14. 11. (v) 1525 Basle: Cartmell Church, 214. (vi) n.d. and p.: Cartmell Church, 257. (vii) n.d. and p.: Hereford Vicars Choral Library, G. 6. 5. (viii) n.d. and p.: Peterborough Cathedral, B. 6. 4. (ix) n.d. and p.: Ushaw College, XVIII. F. 5. 3. (x) n.d. and p.: Wisbech Borough Library, H. 7. 24.

between 1507 and 1512, and between 1526 and 1530, with a completely new set of tools being used, with one exception in each case which forms a link between the two sets, the W.G. stamp between the first and second periods, and the W.G.-I.G. roll dated 1520 between the second and third. This suggests in the former case the death of the father, and in the latter, the death of the elder son, which is confirmed by the appearance in the last period of two panels signed I.G. only,[1] along with the roll bearing the two monograms. The last period of three years, moreover, is marked by a fantastic new style (if anything so unstable can be called a style) in which the two covers differ in design, stamps are used in every way except the way they were meant to be used, and more than once a frame is composed of two different rolls, one for three sides and another for the fourth; this suggests that the elder son, while he lived, was the controlling influence, and that the younger, I.G., was the subject of extreme eccentricity or perhaps inebriety. His life as sole partner was a short, if a merry, one; his bindings cease after three years.

The question remains where this firm worked. I have no doubt at all that the original W.G. was a Cambridge craftsman; two of his stamps[2] are scarcely distinguishable from two used by the Cambridge Unicorn Binder, and were evidently the work of the same die-cutter, while a third[3] was used in conjunction with some of the Unicorn Binder's tools on the work of another binder; the connexion between W.G. and the roll used by L.V.L. and subsequently by Siberch has already been mentioned, and in one case at least W.G. used the Cambridge pink pigment. On the other hand, the tools that appear in the second period strongly suggest what I believe to be the case, that after the father's death, the sons migrated to London; two pairs of panels and three rolls[4] that they used are definitely London ones, while the drawing of the bird and lily, in particular, in the new W.G.-I.G. roll is so similar to that in the rolls used by Reynes as to suggest that they were designed by the same hand.

This brings me to the Unicorn Binder, but so much has been written about him[5] that I do not propose to add much. The number of known examples has risen from forty-three to seventy,[6] and the number of tools he used from twenty-two to thirty-six,[7] since Mr Hobson gave him his name and discussed him in his Sandars Lectures more than twenty years ago. That in the case of a single book, lately come to join the six that have been for some three-and-half centuries in Shrewsbury School Library, out of six stamps that it bears, all but one were hitherto completely unknown,[8] is a fact that gives much food for thought, and its significance will not be lost upon students of bindings. A further remarkable fact is that three of these are close variants of stamps of his already known, and one wonders why. This binder on two occasions resorted to the practice, very rare in England at that period, though not uncommon in Germany, of putting stamps on the back of a book. The only other cases I know are one binding

[1] Ripon Cathedral, XVII. F. 34 (n.p. 1533). There are also rubbings of these two panels in the late Mr Hobson's collection, made by Weale from an unspecified book at Corpus Christi College, Cambridge; unfortunately both books are so worn that it is not possible to identify the subjects.

[2] Nos. 22, 24 and nos. 64, 63. [3] No. 23.

[4] Weale, R. 108, R. 171 and nos. 556, 946, 953.

[5] See Hobson, *Cambridge*, pp. 40–3 and *S.S.L.B.* pp. 52–5.

[6] The following examples of this bindery are additional to those already recorded in Hobson, *Cambridge*, pp. 40–1 and *S.S.L.B.* p. 53: (i) 1482 Padua: Pembroke College, Oxford, 7. d. (ii) 1485 Cologne: Shrewsbury School, R. IX. 53. (iii) 1486 Strassburg: Maggs Bros. Catalogue, 1948, 'Biblia Latina'. (iv) 1487 Venice: J. R. Abbey Collection, Sabellicus. (v) 1487 n.p.: St John's College, Oxford, A. 8. 8. (vi) *c.* 1488 Antwerp: Sotheby's Sale, 29 April 1941, Lot 399. (vii) 1491 Venice: Cartmell Church, Nicolaus Nicolai. (viii) 1491 Venice: National Library of Wales, DR. 5746. (ix) n.d. and p.: Brasenose College, Oxford, MS. 24.

[7] See Pl. XI. [8] The new stamps are nos. 51, 53, 55, 59, 68.

by the Indulgence Binder,[1] and three by the Oxford craftsman whom I call the Fruit and Flower Binder.[2] Sometimes he used pink pigment. It is characteristic of the extraordinary flair that so often led to Mr Hobson's conjectures being proved realities, that his suggestion that the Unicorn Binder may have been Walter Hatley, who appears to have died in 1504, is given added force by the fact that none of these new twenty-eight examples bears a later date, and that in the year after Hatley's death the Unicorn Binder's tools are found being used by other Cambridge binders.

Of the Demon Binder I can say little that is not already known, except that two more bindings[3] of his have come to light, on one of which occurs a new fleur-de-lis stamp which has the peculiar dotted border line that characterises two other of the tools that his die-cutter supplied to him. There are, however, six other binders, in some cases clearly distinguishable, in others less so, who can be added to the list of Cambridge craftsmen, and another who, if he was not a Cambridge man, probably came from the eastern counties.

The most distinctive is the Lattice Binder,[4] whom I have so named after his lattice stamp that was afterwards used by Godfrey, and a border stamp which appears to have been designed to match. On the forty-four bindings I know he used no roll, but nineteen stamps, including six which he acquired from the Unicorn Binder's fount, three of which afterwards passed to Godfrey, and one that later belonged to Spierinck (Gray, no. 2). He seems to have been active from 1485 to 1511. His design was nearly always a frame formed by a rectangular stamp, most often one containing two quatrefoils in lozenges, with the interior divided by diagonal fillets into lozenge and triangular compartments containing respectively a lattice and half-lattice stamp. An interesting variant was using for the frame a half-lattice stamp pointing alternately inwards and outwards. Most of his bindings have been thickly covered with pink pigment.

Another binder, whom on account of the character of his fleur-de-lis, scroll, and eagle stamps I call the Heavy Binder, seems to have worked from 1485 to 1505. I know seventeen of his bindings, and they bear nine different stamps (and again no rolls), one of which, a lattice stamp with sixteen perforations, was used also by the Lattice and Unicorn Binders, while another, the fleur-de-lis, passed into the hands of Spierinck. He, too, used the common English design, but as he had no rectangular stamp for the frame, he generally used a scroll stamp, inscribed IHS MARIA, for the purpose. Some of these scroll stamps, it should be mentioned, are not always easy to distinguish. His most interesting tool was one representing a bird on the back of a quadruped, a subject that occurs on one of the Indulgence Binder's stamps, the origin of which has been discussed by Mr Hobson.[5] The subject seems to have had a wide vogue, for I have found it on two other stamps, one from Polder,[6] and the other French,[7] and in a roll which appears to be Spanish.[8] The Heavy Binder used pink pigment, but only rarely.

The next binder, of whose work I know only five examples, and I am not absolutely positive that they are by the same man, covers the short period from 1480 to 1485. I call him the Pre-Unicorn Binder, because, out of his ten tools, seven were later used

---

[1] St John's College, Oxford, U. 2. 14 (Milan, 1476).
[2] All Souls College, Oxford, SR. 29. f. 4 (Venice, 1491); SR. 30. f. 7 (Venice, 1496); SR. 34. a. 7 (Venice, 1493).
[3] Glasgow University, Bn. 7. b. 4 (Nuremberg, 1481); Christ Church, Oxford, e. 5. 52 (Strassburg, 1491).
[4] See Pl. XI. This binder and the three following are discussed more fully in *S.S.L.B.* pp. 7–12.
[5] Hobson, *Before 1500*, pp. 20–1.
[6] On Victoria and Albert Museum Biblia, 24. i. 65 (Lyons, 1523).
[7] On a French MS. Nocturnale in Sir S. Cockerell's collection.
[8] On Glasgow University, Eg. 6. a. 15 (n.p. 1506).

by the Unicorn Binder, though the bindings bearing them were certainly not the latter's work. One of the others, a fleur-de-lis in a circle, was also used by W.G. Two of his bindings show the feature sometimes met with elsewhere, the use of a small star at each corner of a lozenge stamp. I do not think he ever used pink pigment.

The Monster Binder, named after a strange beast which looks round in some surprise at seeing his tail develop a head at the end of it, is a much more dubious person than the three preceding. The difficulty is with the dates of the books that he bound, and with the fact that only four examples of his work are known to me on which to build conclusions. For out of the thirteen stamps that he used all but one, the monster, were used also by other binders, two by the Lattice Binder, one by the Heavy Binder, and the rest by the Unicorn Binder. One of his books is dated 1505, but the other three 1481, 1495 and 1497, at a time, that is, when the other binders were still working. It is possible, but does not seem probable, that he borrowed tools from his contemporaries. The more likely explanation is that he did not start work till about 1505, and acquired some of the tools of the Unicorn and Heavy Binders when they died or retired about that year, and acquired the two tools of the Lattice Binder during his lifetime. This idea is confirmed by the facts that the Lattice Binder is not known to have used the tools in question after 1505, and that the 1495 and 1505 books are all but identical in their bindings.

There is one book in three volumes sold at Sotheby's in 1941, which does not fit in with any of the foregoing groups, but which must be Cambridge work, not merely because it bears the lattice stamp used by the Lattice Binder and later by Godfrey, but because it is also covered with pink pigment. But the other three stamps[1] on it are quite unknown to me, though similar ones are common (a border stamp of quatrefoils in lozenges, a crested stamp ending in an acorn and flower, and a square with concave serrated sides—not common in England), and, unlike other Cambridge binders, this one used a different design on the two covers. The book is a Gerson[2] printed at Strassburg in 1494, but I can say no more about it, as I cannot trace the book, and I know it only through rubbings sent me by Mr Hobson. But it adds another to the distinguishable binders, of whom I have now named twelve, who I think can be claimed as quite definitely Cambridge men, apart from Z.C. and two whom I am about to mention whose home is perhaps less certain. When there is such general passing to and fro of tools between different binders as there evidently was in the case of the Unicorn, Lattice, Heavy, Pre-Unicorn and Monster Binders, it is reasonable to assume that they all worked in the same town, and there is ample evidence that that town was Cambridge.

Of the two other binders who might be Cambridge men, one I call the Antwerp Binder,[3] because he used a stamp with the arms of Antwerp, and was presumably an immigrant from there. The books he bound are dated between 1485 and 1495, and he used thirteen stamps, the most unusual of which is a square stamp bearing, within an oval, a wingless dragon, holding a flower in its mouth; I know only one example of it, at Caius College, Cambridge,[4] and it is quite unlike any stamp I have ever seen. His design is the normal English one, sometimes with a rosette at the points of intersection of the diagonal fillets. He had a way of often setting a square stamp sideways on, so that the corners pointed up and down. Not unnaturally Weale (R. 374) placed him in Antwerp, but, though he may once have been there, he certainly worked in England.

---

[1] Nos. 132, 133, 134.
[2] Lot 286 in Sotheby's Sale, 29 July 1941.
[3] See Pls. XII and XIV.
[4] Caius College, Cambridge, F. 7. 21 (Basle, 1488).

All the seventeen examples[1] I know have their clasps on the upper cover, six are in Cambridge, one was given to Peterhouse by a member of the College in 1529, one has an English inscription dated 1527, one was bought at the Stourbridge Fair in 1497, one belonged to the Friars Minor of Walsingham, and others belonged at an early date to Coventry School, Durham Cathedral and York Minster, this last having belonged to a monk of Durham College, Oxford, at about the turn of the century. The Stourbridge and Walsingham inscriptions, and the number at Cambridge, distinctly suggest a Cambridge, or at any rate an eastern counties provenance, and there were, of course, many Flemish immigrants living in that district. But it is not possible to be more precise than this.

The other binder is only represented by four examples, all on New College Greek MSS.,[2] and all bound in the Mount Athos style, which was sometimes used by binders, whatever their ordinary style, when they were binding Greek books, namely with a back humped at head and tail, a stud at the bottom edge of the same height as the hump (to make the book stand properly), grooves down the edges of the boards, and straps fastening on to a stud on the edge of the upper cover. On all these four bindings he used the same design, a double frame of stamps, and within it a central foliage ornament surrounded by small stamps, those in the angles being in all cases set askew. All have a number of vertical lines, set very close together, down the backs. The date and provenance of the MSS. are unfortunately unknown, but the bindings are certainly late fifteenth century. This binder, whom I will call the Athos Binder,[3] used eight tools, one of them being the Unicorn Binder's unusual free flower stamp,[4] another being W.G.'s fleur-de-lis,[5] and a third[6] a variant of the rectangular acorn stamp of the Lattice Binder. These facts much suggest that the work was done in Cambridge, though otherwise one would be inclined to attribute to Oxford bindings found only in one of its Colleges, the more so as the design is slightly more Oxford than Cambridge in character, though not quite like the style of either. I have so far failed to identify as Cambridge the work of any later binders apart from those given by Gray.

We must now turn to the productions of Oxford, and here most of the work has been done in Mr Strickland Gibson's pioneer monograph[7] of nearly fifty years ago, though, except for the Rood and Hunt bindings he did not set himself to disentangle and assign to individual workshops the bindings of the fifteenth century. Certain peculiarities strike one about Oxford bindings. One is the survival there, on which both Duff and Hobson remarked, of Romanesque tools and their use again in the fifteenth century; another is the predilection of Oxford binders for intaglio rolls, that is, rolls which, though used 'blind', produce the pattern on the leather in intaglio instead of relief; five of the eleven that I know of as used in England appear to be associated with Oxford. A third peculiarity is the extensive survival into the seventeenth century of roll-decoration, when, except for the Tudor crested ones, the use of rolls had begun to die out in English blind-stamped work. A fourth is the unusually fine leather used at Oxford, which can be the only explanation of the remarkably clear impressions that usually

---

[1] They are: (i) 1485 Strassburg: Caius College, Cambridge, F. 6. 17. (ii) [1486] [Brescia]: Merton College, Oxford, 49. dd. 12. (iii) [1486] n.p.: Peterhouse, Cambridge, O. 6. 14. (iv) 1488 Venice: A. Ehrman Collection, 789. (v) 1488 Basle: Caius College, Cambridge, F. 7. 21. (vi) 1489 Basle: Ripon Cathedral, XVII. F. 10. (vii) 1490 Naples: Westminster Abbey, CC. 30. (viii) 1492 Pavia: Aberdeen University, Inc. 145. (ix, x) 1493 Strassburg: Ushaw College, XVIII. A. 3. 4–5. (xi) 1494 Venice: G. P. Mander Collection, Terence. (xii) 1495 Basle: Christ Church, Oxford, e. 6. 29. (xiii) 1495 Venice: Caius College, Cambridge, F. 9. 18. (xiv) 1495 n.p.: York Minster, XV. A. 12. (xv) n.d. Basle: Durham Cathedral, Inc. 43. (xvi) n.d. Basle: Caius College, Cambridge, F. 6. 4. (xvii) n.d. and p.: York Minster, XIV. B. 22.

[2] New College, Oxford, MSS. 231, 236, 237, 243.   [3] See Pls. XIII and XIV.   [4] No. 131.

[5] No. 129.   [6] No. 127.   [7] S. Gibson, *Early Oxford Bindings*.

survive on Oxford bindings, many of which are still in almost mint condition, when those from Cambridge show the effects of time and wear. It would be unkind to suggest that the reason was that the books were in more use in one University than in the other.

Fifteenth-century binders whom I should be inclined to assign to Oxford can, I must admit, only be so assigned for the most part tentatively. Mr Hobson wrote[1] about the Greyhound Binder[2] who used, besides an unusual triangular greyhound stamp, a circular one of St Sebastian which is probably foreign; it is certainly quite unlike any English stamp I know. Hobson was inclined to make the binder a Londoner, but he knew only two examples. Since then eight others have turned up,[3] though unfortunately only two out of the ten are dated, 1490 and 1496, and I think Oxford is a rather more likely attribution than London. All but two are in Oxford libraries, and of these, six cover MSS. which on the whole are less likely than printed books to have wandered far. One was given to Corpus Christi College, Oxford, by its founder, Foxe, but among the Foxe books at that College are bindings which came from Oxford, Cambridge and London, so this cannot be quoted as strong evidence. This binder, who used fifteen stamps, employed the normal English design (which might be equally Oxford or London), with a frame formed by a rectangular foliaged staff stamp repeated, and usually stamps outside as well as within the frame. His one peculiarity—and it is a remarkable one—was his use of pink plaited headbands, which are very rare in England, and often pink leather thongs for the clasps, which are not very common with calf bindings, though probably they were less uncommon than appears now, because the pink surface may have often been worn off what remains of the thongs.

The Dragon Binder,[4] for whom Mr Hobson did not locate a home, I believe to have been an Oxford craftsman. Mr Hobson knew only two tools that he used, the dragon and a rose, but there are ten, including two rolls, which I should ascribe to him. For he had a peculiarity which is useful for testing his work. His dragon stamp nearly always, and his stag stamp often, were set so as to be seen from the side, not the bottom, of the book. In such cases binders usually placed their stamps so as to be seen on both covers from the fore-edge; not so with the Dragon Binder, who nearly always set them so that, if both covers were turned back and seen together, the stamps are all the same way up. His activities seem to have covered the period from 1486 to 1506, and there are forty-six volumes that I should assign to him.[5] His dragon stamp was damaged, probably in 1505, but was still used by him, and as pointed out earlier,[6] this does help us to decide both the date of certain bindings, and whether some of those with rolls, which he did not ordinarily use, are his work. But I know thirteen other books bearing (among others) his stamps or rolls which I believe to be by a later binder; certainly all that bear his stamps or rolls must not be uncritically attributed to him. One roll[7] that he appears to have used once bears initials that I am ashamed to say I read as A.Q.

---

[1] Hobson, *Before 1500*, p. 24.   [2] See Pl. XV.

[3] The full list is: (i) 1490 Bologna: Merton College, Oxford, 58. b. 9. (ii) 1496 Venice: Worcester Cathedral, Inc. 15. (iii) n.d. and p.: Corpus Christi College, Oxford, Φ. A. 5. 3. (iv) MS.: Pembroke College, Cambridge, MS. 309. (v) MS.: Magdalen College, Oxford, MS. 58. (vi)–(x) MS.: Corpus Christi College, Oxford, MSS. B. 82, C. 31, D. 16, D. 49, D. 58.

[4] See Pl. XV. This binder is discussed more fully in *S.S.L.B.* pp. 46–9.

[5] The following examples are additional to those already recorded in Hobson, *Before 1500*, p. 24 and *S.S.L.B.* pp. 46–7: (i) 1497 [Paris]: Caius College, Cambridge, F. 7. 9. (ii) 1501 Strassburg: Deene Park, II. e. 4. (iii) 1503 n.p.: Stirling's Library, Glasgow, 'Catholicon Januensis'. (iv) n.d. and p.: Ripon Cathedral, XVIII. F. 2. I now regard also as the Dragon Binder's work the nine volumes enumerated, but as only doubtfully his, in *S.S.L.B.* p. 47.

[6] See p. 5.   [7] No. 677.

(which admittedly seemed very improbable) until Sir Ellis Minns courteously drew my attention to the fact that I had read them upside down, and that they were really D.V. joined by a love-knot! It would be tempting to identify these initials with de Villiers who used that very rare thing, a roll with his name in full;[1] the dates correspond, but in spite of the number of tools that each used, no example is known of any of them being used in common, and it is almost certainly merely a coincidence. The attribution of the D.V. roll to the Dragon Binder rests only on one book in two volumes,[2] and that, though dated 1501, must in view of the dragon stamp being in the second state, have been bound a few years later. But, though one of the volumes provides only weak evidence, the conforming of the pattern on the other to the Dragon Binder's peculiar practice makes it certain, to my mind, that it was his work, and D.V. may have been the Dragon Binder's initials. The arguments for assigning him to Oxford are that out of forty-six volumes apparently by him, twenty-four are in Oxford libraries, some of which are known to have been there in the fifteenth century, and that several of his tools were used by another binder with some that were tentatively, and others definitely, attributed by Mr Gibson to Oxford.

Another binder, some of whose stamps appear in Gibson, I call the Fishtail Binder,[3] on account of his most striking stamp bearing an animal whose only recognisable feature is his fish-tail. Of the ten bindings[4] that I should assign to him four are dated between 1473 and 1495, and one 1519, which, in spite of its late date, I think is his work. He used eight stamps, of which one (a very peculiar square one) is used with other stamps on a binding in Mr Ehrman's collection reproduced as Plate I in Hobson's *Thirty Bindings*, which I feel sure is not his work. He has three special characteristics; the two covers are in all cases different, he uses a most unusual design of a frame with a stamp-decorated panel inside, but in all cases but two there is a space above and below the panel left either perfectly plain or only sparsely decorated with three well-spaced stamps, and he sometimes puts three lines down the back. Most of his stamping is exceptionally well and carefully done, though there is one noticeable exception at Worcester Cathedral. Three of his tools are tentatively and one definitely ascribed by Mr Gibson to Oxford, three examples are at Oxford, two of which Mr N. Ker thinks came from Winchcomb in Gloucestershire, where an Oxford binder might well have been employed.

A fourth binder who may have worked at Oxford I call the Floral Binder,[5] because out of his fourteen tools, all but two bear flowers of some sort. His dates appear to be 1477–96, and twenty-two examples of his work are known.[6] Characteristics of him are that the covers were usually different, and that in most cases a larger stamp has smaller

---

[1] No. 956.   [2] Merton College, Oxford, 53. hh. 3-4 (Venice, 1501).   [3] See Pls. XVI and XVIII.
[4] They are: (i) 1473 Louvain: Magdalen College, Oxford, B. 1. 4. 15. (ii) 1481 Oxford: St John's College, Oxford, b. 3. 4. (iii) 1494 n.p.: York Minster, XV. G. 4. (iv) 1495 Nuremberg: Stonyhurst College, 6. 49. (v) [1519] [Venice]: Westminster Abbey, A. 43. (vi) n.d. London: Deene Park, XV. d. 17. (vii) n.d. and p.: Ushaw College, XVIII. c. 3. 5. (viii) MS.: Worcester Cathedral, MA. Q. 73. (ix) MS.: G. Goyder Collection, 'Ambrosiaster'. (x) MS.: Jesus College, Oxford, MS. 102.
[5] See Pl. XVIII.
[6] They are: (i) 1477 Venice: Westminster Abbey, M. 2. 8. (ii) 1486 Basle: Lambeth Palace, 28. 4. 21. (iii) 1487 Alost: Lambeth Palace, 20. 5. 8. (iv) 1487 Venice: Brasenose College, Oxford, UB/S. I. 40. (v) 1488 Basle: Bristol Central Library, EPB. 188. (vi, vii) 1488 Basle: York Minster, XIV. B. 20-1. (viii) 1488 Venice: All Souls College, Oxford, g. 1 infra 7. (ix) 1488 Louvain: Brasenose College, Oxford, UB/S. I. 85. (x) 1488 Basle: Brasenose College, Oxford, UB/S. I. 75. (xi) 1489 Basle: Durham Cathedral, Inc. 44. (xii) 1496 [Nuremberg]: St John's College, Oxford, U. 1. 34. (xiii) n.d. Louvain: All Souls College, Oxford, SR. 27. b. 18. (xiv) n.d. and p.: Worcester Cathedral, Inc. 33. (xv) n.d. and p.: St John's College, Cambridge, Ii. 3. 47. (xvi) n.d. and p.: Ushaw College, XVIII. A. 6. 2. (xvii) n.d. and p.: Lambeth Palace 2. 6. 20. (xviii) MS.: Magdalen College, Oxford, MS. 63. (xix) MS.: Corpus Christi College, Oxford, MS. C. 96. (xx, xxi) MS.: Corpus Christi College, Oxford, MSS. C. 112-13. (xxii) MS.: Corpus Christi College, Oxford, MS. E. 227.

ones—star, rosette or fleur-de-lis—at each corner if it is a lozenge, and round the circumference if it is circular. A small star is sometimes used at the intersections of the fillets, and in that case, and nowhere else, it was apparently always covered with gesso. This use of gesso, to which I shall recur,[1] though not uncommon in Italy, is rare in England where it seems always to be confined to intersection stamps. Although two of the Floral Binder's lily stamps are extremely like those used at Cambridge, of which there are three variants, and although he at least once used lines down the back, it seems more likely that he was an Oxford binder, as twelve out of his books (six of them MSS.) are in Oxford libraries and one stamp is Gibson's stamp 87. The border stamp, moreover, is quite as like some used at Oxford as are the lily stamps like those of Cambridge.

Lastly there is an undoubtedly Oxford craftsman[2] who used a very handsome border stamp of fruit and flowers (we might call him the Fruit and Flower Binder) whose work is common at Oxford, but I have found it only once elsewhere. His dated books range from 1491 to 1509, and seven of his twelve stamps are given by Mr Gibson, while another is the same as Gibson's stamp 86, but reversed. There is an opulent feeling about his bindings, which have usually a double frame, and sometimes a large stamp of a crested character outside the outer frame, and the fine designing and cutting of his tools, which are nearly all large but never clumsy, give an unusually handsome appearance to his work, though there is a sameness about the designs he uses. Curiously enough, with all the self-respect he shows in his bindings, he sometimes allows himself to use his foliage ornament with the end broken off. On three of his bindings he uses a stamp on the back.

There must, of course, be many other Oxford binders represented by the stamps which Mr Gibson reproduces, and very likely others, but either I have not found enough examples to justify the grouping of the bindings as the work of one man, or, if I have, I have lacked the acumen to associate them with Oxford. There are also certainly many stamps which Mr Gibson did not reproduce which probably belong to Oxford, among others two different lattice stamps, each with nine perforations, which are almost exclusively used with Gibson's rolls or stamps, and a third used by D.V.[3]

There are, too, several rolls which, I think, should be added to Mr Gibson's list of those used at Oxford.[4] As rolls are usually quite indescribable, I shall refer to them by the letters and numbers which they bear in my own classified list of rolls. HM. a (5), which appears on books dated between 1532 and 1543, is in seven out of nine cases used with one or another of three rolls which Gibson definitely attributes to Oxford (Gibson, VII, VIII and X),[5] and four of the bindings have the Oxford saltire pattern. HM. d (1), 1530–73, similarly is used with three of Gibson's definitely Oxford rolls (Gibson, XII, XIV and XXI)[6] once each, and also seven times with another roll, FC. c (1), which there is reason for attributing to Oxford. In one case the saltire pattern is used, and one book, at St Andrews, was given by a Magdalen man. Most of the earlier examples, however, show no Oxford association, and the roll may once have been in London. FC. c (1), 1523–57, was used once with two definitely Oxford rolls (Gibson IX and XIV),[7] and often with HM. d (1); three examples have the saltire pattern. MW. d (9), 1528–53, is used four times with a definite Oxford roll (Gibson XXIII),[8] and also with two of the other rolls that I have attributed to Oxford. The design of the roll itself is of a type common at Oxford, and twice the saltire pattern is used. AN. l (1), of which I know only five examples, and only two of them are dated,

[1] See p. 29. [2] See Pls. XVII and XVIII.
[3] Nos. 206, 207, 208. The binders C and M may have been Oxford craftsmen. See Pls. XXVI, XXVII.
[4] See Pl. XIX. [5] Nos. 888, 770, 595. [6] Nos. 857, 861, 728. [7] Nos. 901, 861. [8] No. 829.

1502 and 1505, is used twice with Gibson's stamp 90, and in all cases with one of the stamps I have attributed to Oxford. IN. (3), 1581–1617, is used twice with a definite Oxford roll (Gibson, XXIII),[1] six of the nine examples I know are in Oxford Colleges, and it belongs to a period when ordinary rolls were used little outside Oxford. These reasons may not seem convincing, but it is worth noting that in all cases the Gibson rolls are ones that he assigns without qualification to Oxford, and that none of these new rolls that I have added to the Oxford list has ever been found used with any other rolls or stamps except those given by Mr Gibson and those which I have attributed to Oxford.[2] Mr Gibson refers to but does not reproduce three other rolls: SIHESUS MARIA,[3] which was used with his roll XX,[4] and sometimes in saltire form, but, as he suggests, it was probably used elsewhere besides Oxford; a narrow variant[5] of his musical instruments roll, which is associated with his roll IV;[6] and an S-shaped intaglio roll[7] used with his II, V and VI.[8]

There are five Gibson rolls which must be mentioned, though they cannot be fully discussed, III, IV, V and VI, which he puts down as only '?Oxford', and II, which he hesitatingly ascribes to Oxford, chiefly because one of the two examples he knew was on a Brasenose account book. Rolls III and IV[9] are used together by the same binder, and so are V and VI,[10] so what is said of one of each pair probably applies largely to the other. Mr Neil Ker has found several Oxford printed and MS. fragments in a binding of 1541 at the Bodleian bearing roll V, which is sufficient evidence that it was at one time used in Oxford, and it is likely that rolls III and IV were too; but all four are so mixed up, directly or indirectly, with London, that I am inclined to think that they were more used in London than in Oxford. I should venture to differ from Mr Gibson's attribution to Oxford of roll II,[11] bearing the mysterious letters R.H.M.I. (though it is impossible to say with which letter the inscription is meant to begin); fourteen examples are now known instead of only two, and there seem to be no real links with that town; four of them are blank books which might well be an argument for London, which is known to have supplied blank books all over the country.

London, Oxford and Cambridge were clearly the main centres of bookbinding in the fifteenth and sixteenth centuries, and though decorated bindings are known to have been produced at Canterbury, Salisbury, Old Bokenham, Glastonbury, Jervaulx, and possibly Tavistock, York and Winchester, I do not myself believe that much was done outside those three centres. I can add nothing to what is known about the work produced at these other places,[12] and before leaving the provinces for London, there is only one bindery[13] I want to say a word about, and that is one at Canterbury, which may well have been the monastic bindery of Christ Church, as two of the examples are directly connected with that monastery. Mr Hobson recorded six Canterbury bindings,[14] one of which I cannot trace. But the other five are from the bindery of which I am speaking, and I have been able to add five more.[15] None is dated, but the work is clearly of the fifteenth century. In common with other Canterbury bindings they show the

---

[1] No. 829.

[2] To the above must be added IN (11), for the knowledge of which roll I am indebted to Mr N. Ker, who recently drew my attention to it.   [3] See Gibson, p. 36 and no. 198.   [4] No. 873.

[5] See Gibson, p. 29 and no. 199.   [6] No. 891.

[7] See Gibson, pp. 29, 30 and no. 200.   [8] Nos. 930, 826, 583.

[9] Nos. 951, 891.   [10] Nos. 826, 583.   [11] No. 930.   [12] See Pl. XXVII.

[13] See Pl. XX.   [14] Hobson, *Before 1500*, p. 15.

[15] They are: Magdalen College, Oxford, MS. 88; Canterbury Cathedral, Z. 10. 18; Society of Antiquaries MSS. 47, 287; Sotheby's Sale 22 May 1950, Lot 123*a*. I am indebted to Mr A. Hobson for knowledge of this last.

almost entirely peculiar feature of a chequered pattern. The only use of it outside Canterbury that I know is on one side only of two books[1] by the Scales Binder. Apart from this, the characteristics of this bindery are the two covers being always different, and not only the large fount, but also the large number of tools, in some cases as many as fourteen, used on a single binding. The chequered pattern is produced by the use of a square or a triangular hatched stamp, alternating with a plain space of the same size, and is in the form either of a rectangular or of a circular panel, within an elaborate frame, sometimes including the inscription, TIME DEUM, the whole being often bordered outside by a row of hatched triangles pointing outwards.

[1] St John's College, Cambridge, MS. 23 D; Guildhall, MS. 208.

LECTURE III

OF the bindings that cannot be identified with any provincial centre the vast majority are probably London work, so I propose to take all the rest together, specifying the fact when there is fairly clear evidence connecting them with London. There is time to speak only of a few out of the large number of binderies that it is possible to distinguish, and the still larger number of groups of bindings from which the products of individual workshops have yet to be disengaged, during the hundred and fifty years of which I am treating.

I will mention first two binderies[1] discussed by Mr Hobson, though I can add little to what he said. He pointed out[2] that Caxton appeared to have employed a binder who came from the Netherlands, probably Bruges, and suggested that he was followed soon after 1483 by another who came from South Germany, on the ground that the earlier bindings were characterised by the use of a free fleur-de-lis and a dragon, which were stamps of a type commonly used at Bruges, while in the later ones a griffin stamp, of the type associated with Koberger, appears. This may be true, but exactly the same characteristic dragon frame[3] used in 1483 with a pattern of fleur-de-lis in the panel within,[4] continues to be used with the griffin stamp as late as 1503. Of the eighteen bindings[5] now known to me that I should attribute to this workshop, bearing eighteen different stamps, there are five pairs, each of very similar design, but the various pairs are not much alike, and of the remainder, two (both on books printed by Caxton) are completely different from the rest, one having three vertical rows of stamps,[6] and the other a frame of rectangular stamps with a star stamp at the intersections of the diagonal fillets which divide up the panel within.[7] Unfortunately only nine of the books are dated and I find it very difficult to know where to draw a line between one binder and his successor, if indeed there were two, and not either one or several.

The other craftsman[8] is the one called at first by Mr Hobson the Rebus Binder, who

---

[1] See Pl. XXI.   [2] Hobson, *Before 1500*, pp. 19–20.   [3] See Hobson, *Before 1500*, Pl. 42.

[4] The same pattern of fleurs-de-lis arranged to form crosses is used by Binder K (see Pl. XXVII) on Pembroke College, Cambridge, B. 9. 60 (Florence, 1492).

[5] They are: (i) 1477 Westminster (Caxton): British Museum, IB. 55143. (ii) *c.* 1477 Westminster (Caxton): Pierpont Morgan Library, *History of Jason.* (iii) 1478 Westminster (Caxton): British Museum, IB. 55146. (iv) 1479 Westminster (Caxton): Pierpont Morgan Library, *Cordiale.* (v) 1481 Basle: St Albans School, Z. 1. (vi) 1481 Nuremberg: Pembroke College, Cambridge, C. 17. (vii) 1483 Oxford: Corpus Christi College, Cambridge, Inc. 113. (viii) 1483 Westminster (Caxton): ownership unknown, Gower, *Confessio Amantis.* (ix) *c.* 1487 Westminster (Caxton): Pierpont Morgan Library, *Royal Book.* (x) *c.* 1490 Westminster (Caxton): Baptist College, Bristol: *Myrrour of the World.* (xi) 1495 Westminster: Lincoln Cathedral, Inc. 93. (xii) 1503 Lyons: Corpus Christi College, Oxford, Δ. 10. 14. (xiii) 1504 Strassburg: Cambridge University Library, Rel. c. 50. 1. (xiv) n.d. Westminster (Caxton): Lambeth Palace, 2. 6. 3. (xv) MS.: Chetham's Library, Manchester, Mun. A. 3. 129. (xvi) MS.: Record Office, E. 164/12. (xvii) MS.: College of Arms, C. G. Y. 72. (xviii) MS.: British Museum, Add. MS. 10106.

[6] Pierpont Morgan Library, *History of Jason.*   [7] Pierpont Morgan Library, *The Royal Book.*

[8] See Pl. XXI.

used a stamp with what he subsequently identified as a needle and thread, with a hare and the initials H.C., and he conjectured that the binder's name was H. Cony,[1] working about 1483–92.[2] I only mention him because I do not think that the fact has been published that Mr Hobson's brilliant conjecture was justified, as Mr Graham Pollard drew his attention to a binder of about that period named Henry Cony, who appears in a lawsuit in 1500. I was foolishly incredulous when Mr Hobson propounded his theory, but I was quite wrong; he never, however, persuaded me of the likelihood of this binder being also the owner of a roll[3] with H.C. and a trade-mark, a rabbit, a wivern and a bird, which is found on thirteen bindings, two of 1496 and the rest 1500–15, which I cannot localise at all. Except for the similarity of a rabbit to a hare, they have nothing whatever in common, apart from the initials, with Cony's work, and their designs are much more commonplace. I think more evidence is required to make the identification seem even likely.

A craftsman whom I call the Crucifer Binder[4]—because of his little panel (or stamp), very similar to one illustrated in Gibson—is intriguing because of the mysterious legend beneath it, TR TR ACO MULC. The seated figure, in spite of the absence of a nimbus, must surely be Christ, as he bears a cross over his left shoulder. But in his right hand he appears to hold a scourge, or something similar, and, though I have obtained many opinions on the meaning of the legend, Mr Charles Johnson's interpretation seems to meet with most approval; *Trituracio mulcet*, which may be freely translated, 'Whom the Lord loveth He chasteneth'. Unfortunately, of the five bindings I know,[5] not all of which bear this stamp, but which I believe to be all by the same binder, only one covers a book which can be dated, 1476, the others being manuscripts which are attributed to Norwich, Canterbury and Rouen, a fact which does not help much in localising the binder. In a collection of Weale's rubbings that belonged to Mr Hobson, there is an example of two out of this binder's fourteen stamps, and they are used with Pynson's rose panel, which brings in London. The Crucifer Binder sometimes used metal bosses (not cornerpieces) in the corners of his bindings—not very common in England, though there can, I think, be no question about his being an English binder.

The Bat Binder[6] is represented, as far as my knowledge goes, by only three examples,[7] dated 1486–9. I name him after his characteristic tool, a bat fitted into a half-circle, which he used doubled, back to back, as a frame. His designs, in two cases different on the two covers, are unusually striking, as is also his fount, for it includes among its eight tools, three, exceptionally well-cut, representing saints, of a type much more common abroad than in this country, St Lawrence, St Catharine and one who may be St Barbara. But he is certainly English; the clasps are on the upper cover, the Eton

---

[1] Hobson, *Abbey*, p. 185.

[2] One example can be added to Hobson's list: Holkham, Anselm (Antwerp, ?1491).   [3] No. 566.

[4] See Pl. XXI. Mr K. Harrison, of King's College, Cambridge, has suggested to me that the small stamp bearing flowers in a vase of classical shape may be a representation of a Garden of Adonis, and that the larger stamp shown on the binding reproduced by Hobson (*Before 1500*, Pl. 34) of a MS. written for Gilbert Kymer, Dean of Salisbury, is even more likely to have been intended as such. For Kymer was physician to Humphrey, Duke of Gloucester, and made use of one of his mottoes, and Mr Harrison suggests that he is therefore quite likely to have had the cover of his MS. decorated with a stamp of a Garden of Adonis, which has been shown by Sir T. D. Kendrick (*Antiquaries Journal*, vol. XXVI, pp. 118 et seq.) to have been a badge which Duke Humphrey used, and which decorated his tomb.

[5] British Museum, Add. MS. 22573; Fitzwilliam Museum, MS. 375; Magdalen College, Oxford, MS. 166; Chetham's Library, Manchester, Mun. A. 2. 171; St Albans School, Y. 1. (n.p. and d).

[6] See Pls. XXII and XXIII.

[7] They are: (i) 1486 Basle: York Minster, IX. G. 11. (ii) 1489 n.p.: St Albans School, Z. 2. (iii) MS.: Eton College, MS. 119.

MS. has a complete English pedigree,[1] and the St Albans example has on the fly-leaf an early English note of ownership. Very likely the tools are foreign importations. In one case he used vertical lines down the back.

There is a binder whom I call the Lily Binder,[2] on account of his predilection for lily stamps of which he had two variants, one or other of which appears on practically all his bindings. One of them is attributed by Weale-Taylor (no. 313) to Germany, but whether this binder ever worked in Germany, which I doubt, he certainly worked in England. I have hesitated to attribute to him all the bindings bearing stamps which he used, but it is hard to reach any other conclusion than that they were all his work. The difficulty about him is that the bindings bearing stamps from this group have a number of distinctive features, but each of them appears on only some of the bindings, but they all, so to speak, overlap, and it is impossible to arrange the books in distinct groups on the basis of these features. They are: reddish leather, sharply bevelled boards, a double frame decorated with alternate round and lozenge stamps, pink plaited leather headbands, yellow edges, metal shoes on the edges of the boards, and, very odd, gesso used on small stamps only where the fillets intersect and always on one cover only, there being no corresponding stamp at all on the other cover. Exactly the same curious limited use of gesso occurs on one binding of the Pre-Unicorn Binder at Wells,[3] but no other example is known to me. Five volumes, four[4] of which are parts of the same book in Cambridge University Library, have clasps on the lower cover. I surmise, from the first four and last-named features that the binder was an immigrant from the Netherlands, and retained some of his native characteristics. It might be thought that some of the bindings were executed by him before he came to England, but that theory, though it may be true, does not work out easily. If I am right in attributing to him all the bindings in question, there are eighteen of them,[5] ranging from 1481 to 1504, bearing twenty-four different stamps.

The books bound by the Half-Stamp Binder[6] are dated 1491–1511, and number twenty-two, on which eleven different stamps and a small panel are used. The first peculiarity one notices is that the lattice ornament in the lozenge compartments is not a single stamp, but two halves put together. Usually the ornament is divided horizontally, but, curiously enough, this binder also used (on a book at Lincoln)[7] a similarly composite ornament divided vertically. I know one other binder who thus used half-stamps, but he used rolls, and was active from 1515 to 1530. The converse, so to speak, is sometimes done, as it is possible, by 'digging in', to use a tool with the whole ornament, but impress only half—a practice which is usually betrayed by inaccurate impressing. Most of the Half-Stamp Binder's work is rather uninteresting, but some of his stamps—not the half-stamps—occur on two bindings of much interest; one is the empty covers at Westminster Abbey which bear a small Royal Arms panel (perhaps, after all, in view of Mr Nixon's discovery of the book belonging to it, the earliest panel, as once claimed by Duff, used in England), and an azured heart and an outline hand stamp;

[1] Information from Mr N. Ker.　[2] See Pl. XXIII.　[3] Wells Cathedral, B. 1. 36 (Cologne, 1485).
[4] Cambridge University Library, Justinian 1543, 1546–8.
[5] They are: (i) 1481 Nuremberg: York Minster, X. A. 7. (ii) 1481 Nuremberg: Corpus Christi College, Oxford, Φ. B. 3. 2. (iii) 1483 n.p.: Ripon Cathedral, XVII. F. 7. (iv) 1486 Ulm: Hereford Cathedral, L. VII. 13. (v) 1487 Nuremberg: Christ Church, Oxford, Allestree, J. 1. 19. (vi) 1488 Tubingen: Corpus Christi College, Oxford, Φ. E. 2. 9. (vii) 1498 Basle: Caius College, Cambridge, F. 6. 26. (viii–xi) 1499 Venice: Cambridge University Library, Justinian 1543, 1546–8. (xii) 1503 Paris: Corpus Christi College, Oxford, Δ. 6. 13. (xiii) 1504 Paris: Peterhouse, Cambridge, O. 8. 12. (xiv) n.d. Louvain: Oriel College, Oxford, W. 9. 1. (xv) n.d. and p.: York Minster, XIX. C. 5. (xvi–xviii) n.d. and p.: Christ Church, Oxford, Allestree, A. 3. 1–3.
[6] See Pls. XXIV and XXV.　[7] Lincoln Cathedral, SS. 1. 16 (Basle, 1493).

the other is an Augustine[1] belonging to Major Abbey, which has the hand, but not the heart stamp. When I wrote previously on this group of bindings[2] I hesitatingly suggested that, though so different, the two more interesting examples might be regarded as probably by the same binder. But since then, the discovery of thirteen more examples[3] has made clear a peculiarity of this binder that appears in all the group, the setting of the stamps in the frame upside down on one side and facing different ways at top and bottom, so that they have the effect of appearing to march round the centre panel. As this is very unusual, and, as far as I know, never systematically done by any other binder, I have now no hesitation in claiming Major Abbey's binding and the Westminster one as the work of the Half-Stamp Binder.

A man whom I have called the Octagonal Rose Binder[4] after a stamp that he used of unusual shape, whose books cover the short period 1489–96, often used the also rather unusual shaped stamp of a flower in a triangle which Mr Gibson attributed to Oxford. But probably the tool belonged to this binder first, for there is nothing particular to connect him with Oxford. On the ten examples I have recorded[5] he used ten stamps, and his only peculiarity—and it is not unique, though unusual—is placing a group of three rosettes where the bands join the sides.

Another binder[6] of this period deserves mention, on account of one beautifully cut tool that he used representing a huntsman, and one triangular dragon stamp of his which is remarkable because it has to be looked at with the apex downwards. There are no peculiarities in his work, which makes it extremely difficult to know which of the bindings bearing his stamps are by him, as some of them are linked up, directly or indirectly, with a bewildering number of other bindings, some bearing initials, and one the mysterious inscription A BATAS. I should be inclined to assign to the Huntsman Binder only ten books[7] that I know; if that is correct, only ten stamps (one undecipherable) are his, and his books are dated between 1477 and 1498.

The last of the fifteenth-century binders that I will mention I call the Foliaged Staff Binder,[8] because on all the bindings that are likely to be his there is used in the frame a rather crude stamp of a ragged staff with foliage. It is possible that three bindings[9] bearing this stamp, two of which are on Greek MSS. bound in the Mount Athos style, are not his work, as, apart from this difference (which is by no means conclusive by itself of the work being that of a different binder) there are other distinctions. On one of them are used a triple panel of animals in foliage and oddly enough, for one small section only of the frame, a different ragged staff stamp; neither panel nor stamp is otherwise known to me. On all three the stamps are carried straight through the

---

[1] Hobson, *Abbey*, Pl. 2.　　　[2] *S.S.L.B.* pp. 36–8, where a list of these bindings is given.

[3] They are: (i) 1489 Rome: St John's College, Oxford, A. 1. 10. (ii) 1498 n.p.: Harsnett Library, Colchester, K. F. 11. (iii–ix) 1502 Basle: Ushaw College, XVIII. B. 3. 5–11. (x) 1504 Venice: Wisbech Borough Library, D. 5. 10. (xi) 1505 n.p.: Ushaw College, XVIII. A. 5. 4. (xii) 1511 Paris: Pembroke College, Cambridge, D. 10. (xiii) n.d. and p.: St John's College, Oxford, E. 4. 50.

[4] See Pl. XXV.

[5] The list is given in *S.S.L.B.* p. 56, to which must be added Cambridge University Library, Rel. B. 50. 1 (n.d. and p.) and Worcester College, Oxford, P. 1. 15 (n.d. and p.).

[6] See Pl. XXV.

[7] They are: (i) ? 1477 Cologne: Pembroke College, Cambridge, C. 15. (ii–iii) 1486 Venice: Lambeth Palace, 7. 5. 4–5. (iv) 1490 Venice: Corpus Christi College, Cambridge, 57. (v) 1497 n.p.: Deene Park, V. d. 7. (vi) 1497 Lyons: Emmanuel College, Cambridge, MSS. 5. 2. 2. (vii) 1498 Nuremberg: New College, Oxford, Auct. Wagon, X. 11. (viii) 1498 Strassburg: Corpus Christi College, Oxford, Φ. E. 2. 8. (ix) n.d. and p.: Edinburgh University, Inc. 122. (x) MS.: Trinity College, Cambridge, MS. 1182. The huntsman stamp is discussed in Hobson, *Cambridge*, pp. 45–6.　　　[8] See Pl. XXVI.

[9] New College, Oxford, MSS. 232 and 244; Corpus Christi College, Oxford, MS. 23.

points of intersection of the frame, and no small stamps appear outside it, whereas the normal design of the other bindings is to leave the point of intersection blank, or to fill it with a stamp different from that of the frame, and to place occasional stamps between the frame and the edge of the book. If we do include these three, there are thirty-eight bindings, bearing different stamps, that may be attributed to this craftsman, covering books dated 1489–1502.[1]

In dealing even with fifteenth-century binders it has been necessary to be selective, and this necessity is even greater in passing on to the roll period. It will, therefore, only be possible to speak of a few of the sixteenth-century binders whose work can be identified and recognised. In some ways, however, these binders' work is less easy to identify than that of their predecessors, for the use of rolls considerably limits the range of personal peculiarities in design that find more scope in the building of a pattern with stamps.

One cannot omit John Reynes' bindery, the most prolific of the period, although not a great deal new can be said about it. One remarkable binding of his, however, has never been noticed. It covers an official book in manuscript that belonged to Sir Charles Somerset as Chamberlain to Henry VII, and it is in the collection of Mr J. W. Hely-Hutchinson. It is unique among Reynes' bindings in four respects: it bears three panels, the Baptism (first state), the St George, and the animals in foliage (Hobson, *Panels*, Pl. 4); it has a flap over the fore-edge; its boards consist of leather—a feature rare at any time—taken from an old German binding; and, if it was bound as a blank book, which is probable since Somerset was only appointed Chamberlain as late as 1508, quite a likely date for this binding, it is the only known example of Reynes binding a blank book.

Mention has already been made[2] of Reynes' two similar rolls, the unsigned and the signed, and we are lucky because the man who used his signed roll after his death in 1544 erased his initials, and saved us the confusion created by so many binders who used other people's signed tools. It is, I think, therefore, safe to attribute to the Reynes' bindery any book that bears the signed roll. There can thus be ascribed to him a conventional foliage roll[3] and a roll with a fleur-de-lis and half daisies,[4] and, of stamps, the ordinary pineapple[5] (omitting the corresponding half-stamps throughout), a large intaglio ornament,[6] and two stamps which I have only found used by him on one book each, a cruciform ornament,[7] which was used by one or more other binders, and an ornament with a flower in the centre,[8] which I do not know otherwise. His panels, of course, I am leaving out of consideration. Three other rolls are perhaps only indirectly connected with him; a floral roll[9] which was very often used with two panels[10] that once belonged to him, but probably before they came into his possession, and which was

---

[1] The stamps used by fifteen other early binders, who are distinguished only by letters, are reproduced on Pls. XXVI and XXVII, in case, though I can give no information about them, and in some cases only a single example is at present known, it may be possible to relate to them other bindings that come to light. The dates of the books on which they appear are:

Binder A.	? 1480–8	Binder I. (Salisbury)	MS.
Binder B.	1481	Binder J. (? Salisbury)	MS.
Binder C. (? Oxford)	1481–3	Binder K.	MS.
Binder D.	1483–6	Binder L.	MS.
Binder E.	1486–91	Binder M. (? Oxford)	1506
Binder F.	1488–95	Binder N. (Old Bokenham)	1519
Binder G.	1492	Binder O. (Glastonbury)	c. 1530
Binder H.	1497–1507		

[2] See p. 5.   [3] No. 434.   [4] No. 435.   [5] No. 437.
[6] No. 438.   [7] No. 440.   [8] No. 439, used on Stonyhurst College, 1. 9 (Paris, 1516).
[9] No. 436.   [10] Weale, R. 109.

also used with a bewildering number of rolls, stamps and panels; Siberch's dancing peasants' roll[1] which was used with some of Reynes' stamps and possibly by him, before the year 1522 when it is found with Siberch's signed roll; and the extremely interesting roll of unusual type, bearing St Barbara, Adam and Eve, and the name of de Villiers,[2] which is used with Reynes' pineapple on two out of the only eight books (in twenty-two volumes) on which it appears. Both this and the floral roll are linked up with that elusive group which includes some of Gibson's rolls which I have mentioned before[3] as appearing to be connected with both Oxford and London, and the latter also with the apparently Oxford G.P. panels.[4] I only wish I could bring order out of this confusion, and distinguish the work of one binder among these rolls, but at present it seems hopeless. I doubt, however, whether these two rolls were ever actually used by Reynes.[5]

The binder F.D.,[6] whose signed roll appears on thirty-five bindings which date mainly from 1554 to 1580, is easy to recognise through his work, and is of some interest. Four examples with his signed roll must almost certainly be by a successor, as not only do they bear rolls which he does not seem to have used, but they differ from his design, which is quite unmistakable. For he appears to have used only two rolls, one with the heads of Erasmus, Huss, Luther and Melanchthon, with his initials on either side of the last-named head, and a heavy quatrefoil diaper roll. And his design was, with very few exceptions, always the same, one roll, used doubled, for a frame, with an inner frame touching the outer at the sides, but leaving a plain space at top and bottom; the centre panel is divided into three by horizontal fillets, and horizontal fillets join the bands to the outer frame. In other words, he uses the typical London design, with the variation, as far as I know peculiar to him in England (except for one single example[7] of another London binder), of the plain space above and below the inner frame. Now this is precisely the typical German design, but used as a rule with panel stamps, and identical heads with only slightly varying foliage in between are found on German rolls. He must clearly have been a German immigrant, and as the initials seem distinctly crowded, it is possible that he added them later to a German roll. Occasionally he only uses the diaper roll, but the design still makes these bindings quite easy to recognise.

A familiar London roll is signed H.R.,[8] with, in the other compartments, Tudor emblems. There are three variants of this roll, and of one of them[9] I know only two examples, one at Shrewsbury School[10] and the other belonging to the Oxford University Press to which it was given by Mr Gibson. The remarkable thing about these two examples is that one book is dated 1551 (which fits in with most of the bindings bearing other H.R. variants), but the other is dated 1654, and is not an *emboîtage*. There are also two I.R. rolls, a broad and a narrow, and two rolls, a broad and narrow, with exactly the same monogram on each, which appears to be I.R.T. The broad I.R.T. roll, with dates 1538–61, is used with the narrow I.R.T. and with both the I.R. rolls; clearly therefore, they all belong to the same group. Fortunately the Muniment Room at Winchester College enables us to date some of the bindings with exactness, but the matter is not quite so simple as in a short sketch I shall have to try to make it appear

---

[1] No. 431.   [2] No. 956.   [3] p. 24.   [4] Goldschmidt, no. 169.
[5] Reynes is discussed more fully in *S.S.L.B.* pp. 24–7.   [6] See Pl. XXIX.
[7] Aberdeen University, II. f. 2231 Mer (Geneva, 1573) on which rolls CH. *a* (3), DI. *d* (1) and FC. *d* (1) are used.
[8] See Pl. XXX. These bindings are discussed more fully in *S.S.L.B.* pp. 12–17.   [9] No. 758.
[10] Shrewsbury School, A. III. 10 (London, 1551).

to be. Now the H.R. rolls, used often with some other unsigned ones, seem to be used from about 1550 and they cease about 1584, but the unsigned ones continue. About 1603 the narrow I.R.T. roll appears, but in a very unusual design, with the centre panel divided by several horizontal strips of rolls, sometimes doubled, into as many as six (instead of the usual three) sections. About 1628 this roll ceases, and one of the H.R. rolls reappears but sometimes in the strange design just described. There can be no doubt that I.R. and I.R.T. stand for the same name, though I can make no suggestion why the 'T' was added. That being so, I think the explanation clearly is that we have the work of four generations of one family, extending over some 120 years. The original I.R.—I.R. senior we will call him—worked about 1538-61; his son, H.R., took on the business and died about 1584; his son, I.R. junior, used his father's unsigned rolls, but about 1603 either took over a narrow I.R.T. roll belonging to his grandfather, or had it made for himself on the model of his grandfather's broader roll. He developed a strange style of his own, and died about 1628, when his son, H.R. junior, took over, continuing his father's peculiar style, but using again his grandfather's signed rolls, which bore the right initials. This would explain the extraordinary reappearance, with which I began, of a roll which had been used a century earlier. One, if not both, the I.R. rolls was probably used in both periods; all the H.R. rolls and the broad I.R.T. roll were made in the earlier period, and the narrow I.R.T. roll may have been. The number of examples of all put together that I know is 106. The design used clearly indicates London work.

A prolific binder[1] who signed many, but not all, of his bindings, had the initials R.B. and was certainly a Londoner; fifty-four out of his 160 odd bindings that I attribute to him have the characteristic London lozenge panel, one-sixth of all the books he is known to have bound were printed in London, and three are blank books in different parts of the country, and likely to have been bound in the capital. His most common rolls were his signed roll with heads in medallions, and a roll bearing Tudor emblems (which must be carefully distinguished from another,[2] almost identical, which had nothing to do with him); but he often used a roll with the golden fleece and a falcon, and, as practically all the books bearing this roll fall within the apparent period of R.B.'s activity—to judge by the dates of the signed bindings about 1550–80—they are almost certainly his work. He appears to have used, including the foregoing, seven rolls in all, and possibly an eighth, but another seven were used with his rolls, as I think, by a successor.

The last binder[3] I will mention is the owner of a very handsome roll, a broad one with the Tudor arms and emblems and a curious binder's mark. An odd fact is noticeable at once about its use. It is used on thirty-seven printed books down to 1559, but it is also used on as many as twelve blank note-books, most of which (and the others might have been re-bound) appear to have been written in subsequently to that year. This suggests that the roll (and a close variant of it)[4] passed from the hands of an ordinary binder into those of a firm that specialised in blank books. But there is another interesting fact about this, namely, that there exist, used exclusively on blank books and scattered all over the country, five other rolls[5] which are exactly the same as the one of which I have been speaking, except that, in place of the binder's mark, they have

---

[1] See Pl. XXX. This binder is discussed more fully in *S.S.L.B.* pp. 74-6.
[2] No. 752.
[3] See Pls. XXX and XXXI. These bindings are discussed more fully in *S.S.L.B.* pp. 28–31.
[4] No. 745.   [5] Nos. 738, 739, 736, 740, 744.

respectively the initials I.H. (seven examples),[1] I.S. (two examples), E.P. (two examples), G.D. (one example)[2] and L.P. (one example), and no doubt there are many more to be found. The obvious explanation of initials on rolls, that they were those of the head of the firm, is out of the question here; the coincidence is too remarkable that these signed rolls occur only on blank books, and that they are all exactly like one another, and like that used by the Blank Book Binder. The only explanation that I can see is that this firm that specialised in blank books employed a number of craftsmen, and to each was given his own roll of the same design as the firm had long been using, but with his initials added. Such a practice is paralleled by the case of the Dulmen monastic binders, of whom at least six signed their own bindings, only they used stamps, not rolls.

I have now dealt with thirty-eight specific binderies, and there is not time to discuss any of the many others. But there are a few problems that the student of early English bindings meets, to most of which I, for one, can find no convincing solution, but which should be referred to briefly.

What is the origin of the ornaments used by binders? This is a matter of iconography of which I have not sufficiently wide experience. Of course it is known that rolls were frequently copied from woodcut borders in Paris Books of Hours, and sometimes very exactly. But there are ornaments used in these, and also on rolls which have no obvious source, that appear to be quite meaningless, though often they reappear over and over again. Most of these I can only assume are corruptions, due to ignorant designers constantly copying one another unintelligently, of something that once had a meaning. I suggest very tentatively a few possible origins for such ornaments.[3] The three heads side by side, used on scores of rolls and in many woodcut borders, might be derived from heads forming a capital to clustered columns seen from in front. The strange waisted ornament with diagonal stripes and dots between them that occurs on a number of rolls, might be a sort of telescoped form of a twisted column that appears on book borders. The type of roll, used in both England and France, with a twisted stem and pineapple ornament, might be derived from an ornament used in stone. The cherub whose neck is prolonged and divided so as to curl back on either side might have its origin in another stone design where the unrealistic treatment had more artistic meaning. The curious feature of shallow ellipses on the top of one another (sometimes there is only one) might be a reminiscence of a rather similar feature in metalwork, while there can be no question that one type of ornament that is always occurring on rolls is definitely copied from an iron grille, with its 'buckles' holding together the curved members. A possible explanation of the chequered pattern, practically peculiar to Canterbury, might be found in the pattern of split flint and stone that is a feature of a number of walls in and around Canterbury. No such work of the fifteenth century seems to be known to survive, but it is conceivable that the sixteenth-century walls continue a local tradition, which might have been the source of this peculiarity of

---

[1] Examples not hitherto recorded are: Folger Library, Household Accounts of Sir T. Pope (? 1628) and Messrs Sprott, Stokes and Turnbull, Shrewsbury; Wem Court Roll (? 1588).

[2] Folger Library, Household Accounts of Sir T. Pope (? 1628). The occurrence of both the I.H. and the G.D. rolls on this book is not easy to explain on the basis of the theory here suggested; but it is quite incompatible with the only other explanations that have been put forward, namely, that either all these sets of initials represented distinct blank-book binderies patronising the same die-cutter, or they were the initials of different booksellers for whom the same binder worked. There is every reason for thinking that at this time specialisation as between blank-book and letterpress binders was not general, for the large majority of rolls found on blank books were also regularly used on letterpress books. It is, therefore, unlikely that these initials were those of different binderies all of which specialised in blank-book work.

[3] See Pl. XXXII.

Canterbury bindings. Some of these suggestions may be thought fantastic, and there are many ornaments for which I cannot suggest any explanation whatever, but it would be an interesting subject for research.

Who were the designers and the die-cutters who produced the tools that the binders used? Perhaps at the time we are speaking of the two were the same, but wide enquiries have led me to the conclusion that no one knows very much about them. An obvious guess is that the die-cutters were the same as the seal engravers, as the technique must be the same. If so, it is significant that the best engraved binders' tools used in England coincide with the period when English seal engravers had a deservedly high reputation on the Continent, at the end of the fifteenth century.

There are two possible views as to where the binders got their tools from. One is that they were all bought at the great book fairs which enjoyed international patronage, and this would account for the extraordinary similarity of some tools used in several different countries.[1] The other is that they were bought more or less locally, which would explain certain types of roll, and still more of stamps, appearing to be characteristic of particular districts. I feel sure that the first view, put in its extreme form, is untenable, for the similarity of tools locally used is often too marked,[2] but it may well be accepted that some of the rolls that appear with little variation (those with heads in medallions are an obvious example) in two or more countries were produced centrally and imported. I suggest that, in general, the worst and the best engraved tools were likely to be local productions; the worst, because, though they might be the best that a local craftsman could produce, they would not be likely to stand up to international competition, nor would binders be likely to go far afield to acquire them; the best, both because some of them, for instance some of those used at Cambridge and London, are so outstanding that they do not seem likely to be the result of mass production, and because to some extent their use seems to have been confined to particular binders, the Unicorn Binder, for example.

But though a central supply might largely explain similar designs being used in different countries, the possibility of conscious copying to the order of one binder who had admired a tool used by another must be taken into account also. For instance, Spierinck used a beautiful roll with flowers and birds, and, though it is not a common theme, variants[3] of it were used in Italy, the Netherlands, and different parts of Germany, so close that they obviously have a common source, but so different in artistic quality that it is difficult to believe that they could be the work of the same hand. Deliberate copying I should believe to be the explanation of a phenomenon one sometimes finds, namely, rolls which are almost identical but in different sizes;[4] in no such cases do I remember their being used by the same binder. There are other cases of rolls which differ only minutely, and the explanation of them is anybody's guess. My own would be that they were usually the work of the same die-cutter, whether a local man, as I should think in the case of the dated and undated W.G.-I.G. rolls, or one who produced for the international fairs, as might be the case with the boring series of heads-

---

[1] See Pl. XXXIII.

[2] It may be more than a coincidence that a milled bounding line, which is exceptional elsewhere, appears on six tools used at Cambridge; W.G.'s signed and fleur-de-lis stamps (nos. 20, 23), the Lattice Binder's quatrefoil stamp (no. 86), the Heavy Binder's stag stamp (no. 91), the Demon Binder's dromedary stamp (no. 114) and Godfrey's diaper roll (no. 593). A bounding line broken into dots is used on the Demon Binder's goat, wingless dragon and fleur-de-lis stamps (nos. 112, 118, 123).

For close variants used in England, see Pl. XXXIV.

[3] Nos. 479-85.   [4] See Pl. XXXV.

in-medallion rolls used in England (of which I know some ninety different ones), France and Germany. In the German rolls, whose foliage has some merit of design, the heads must have become standardised, for precise copies are produced over and over again. The fact remains that, while some originality is shown in the design of stamps, there are very few types of rolls used in England, except of course the heraldic ones, that have not their counterpart in use abroad, though some differ in small characteristics.

But we are still left wondering who the men were who designed or cut the tools, or perhaps did both, and what their status was. I believe they must have varied indefinitely between the real artist and the ignorant and crudely unimaginative workman. It is, I imagine, very rare to be able to attribute a binder's roll to a well-known artist, but it is possible with regard to two rolls[1] used on a book at York[2] which are beyond doubt the work of Hans Holbein, one of them containing a self-portrait with the initials H.H. These initials were taken by Haebler to refer to Hans Huss, because the other heads are those of Erasmus, Luther and Melanchthon. But on rolls Huss's name always appears in the Johannes form, never as Hans, and in the case of the other heads the names (in shortened form) are inscribed round the edge of the medallions, whereas in this case only initials are used, and they are put in a different position, either side of the head—an obviously intentional distinction; moreover, the portrait is not in the least like Huss and is exactly like Holbein (who did also sign a book border with his initials), and the other portraits are far superior to the heads of Reformers that so frequently occur on rolls. And the matter was put beyond doubt when Mr Charles Bell discovered for me that the companion roll, representing dancing peasants, strikingly more pictorial and vigorous than other examples of the same subject, proved to be only a slight variant of one of Holbein's known woodcuts.

Some of the stamps used, for instance several of the Unicorn Binder's, are beautifully drawn and finely cut, and must be the work of someone very different from the people responsible for many of the rolls that have no artistic claims.[3] But at the other end of the scale there are endless utterly uninspired rolls, often vain repetitions or a series of meaningless and not beautiful ornaments. There is even one panel with an inscription in ornate Gothic lettering running round it, in which the die-cutter, in copying his model, has not only made the not uncommon mistake of producing the result of looking-glass writing by forgetting to reverse it, but was evidently so illiterate that he thought an 'A' exactly in the middle of the inscription on one side was a piece of decoration, and accordingly inserted it to match in the same place on the other side. So wide a difference between the best and the worst produced tools would not be so likely to occur if they were the result of competitive production at an international fair as if they were the work of local craftsmen of varying ability.

What is the significance of initials on a binding? In theory at least they may be those of the binder himself, the stationer who employed him, the die-cutter, the owner, or someone, such as the sovereign, who has no connexion with the book at all. Among English blind bindings there may be a few cases of the last, but very few; I have in mind an unusual roll[4] at the Record Office[5] in which H.R. might stand for Henricus Rex, and the roll[6] with the Bridge House mark on it, an 'S', the Stuart arms and Tudor emblems and a unicorn—all pretty mysterious—and the initials I.R. which may be for Iacobus Rex.

[1] Nos. 521, 527.　　[2] York Minster, VII. B. 3 (Basle, 1532).　　[3] See Pl. XXXVI.
[4] No. 670.　　[5] Record Office, E. 315/251.　　[6] No. 743.

That the initials on English rolls are those of the die-cutter can, I believe, be dismissed. Such are, of course, common in Germany, and Haebler worked out rules for distinguishing them, by their form, from those of the binder, but in these instances there are generally two sets of initials. In England, however, in the few cases where this occurs, for instance W.G. and I.G.,[1] B.B. and N.D.,[2] the two sets of initials take the same form, and look therefore like those of equal partners. If one set alone appears, it seems unlikely that the binder would allow a die-cutter's signature on a book on which he had not placed his own, seeing that everything goes to suggest that, unlike the case of Germany, it was not the regular practice of English die-cutters to sign their tools. I happen to know only three cases (though it may have been commoner abroad than I have been aware) where initials appear on metal cornerpieces, but only one of these is an English one. The binding[3] bears the Bridge House roll, and the letters, R.B., on the cornerpieces are so minute that Mr H. M. Nixon's suggestion seems quite probable that here they are the initials of the maker of the metal fittings. This may apply to the more prominent monogram J.C.[4] which appears on two different sets of cornerpieces that I know used by a German binder, where they are not the owner's initials, which appear elsewhere. But in this case, and in that of another German book with N.M. on the cornerpieces, they may indicate the binder.

Owners' initials, habitual in Germany in the plain space above the panel, are usually in England easily recognisable by being put nearly always either separately on either side of a centre ornament or outside the frame, or else in a small shield-like frame in the centre of the cover.[5] In both cases, each initial is stamped separately. But there is a type of stamp bearing a monogram on a shield, the whole forming one stamp, which can be proved to denote the owner, as in the only two examples[6] I know, I.G. at Hereford and T.G. at St John's, Oxford, the initials are those of the original owner as shown by inscriptions in the books.[7] I can quote also two cases of rolls bearing a coat of arms and initials. One is Godfrey's roll,[8] the G.G. on which, Weale, surely rather improbably, interpreted tentatively as the initials of Guido Gimpus on account of the arms, though the initials are exactly in the form used by Godfrey. The other is a roll[9] bearing a coat of arms and the initials R.F., used four times with one of the I.R. rolls and once alone. The coat appears to be that of Fareley, and the initials may indicate a member of the family for whom the books were bound. The stamp[10] bearing the Percy crescent, found on two books,[11] presumably also indicates ownership.

There remains to be considered the craftsman himself who bound the book and the stationer who employed him, whether as an assistant in his firm or as an outsider. It has generally been assumed that by the 'binder' one meant the actual man who handled the tools and that initials on bindings normally referred to him. This, as I have said, I believe to be the case with the rolls used by the Blank Book Binder. It seems likely,

---

[1] Nos. 559, 560.   [2] No. 700.   [3] Shrewsbury School, R. IX. 72 (London, 1623).

[4] Shrewsbury School, A. IV. 30 (Delft, 1582) and G. D. Lane Collection, *Proverbs of Solomon* (n.p. and d.).

[5] See no. 545.

[6] Hereford Cathedral, Vicars Choral Library, G. 7. 2 (Strassburg, 1559) and St John's College, Oxford, U. 2. 16 (Lyons, 1505); see nos. 546, 550.

[7] The stamps of two owners, not in the form of shields, are reproduced on Pl. LXI; one owner was John Bramfield, whose stamp appears on St Andrews University, Dc. 1. 14 (Paris, 1528), which contains his signature; the other was Ralph Freeman, who gave a number of books to All Souls College, Oxford, c. 1772, on which a device with an F appears, as Mr N. Ker has pointed out to me, in three different forms, one of which (no. 1089) used on sixteenth- and seventeenth-century books, probably belonged to an ancestor, though the other two variants must be eighteenth century.

[8] No. 746.   [9] No. 763.   [10] No. 1047.

[11] Durham Cathedral, N. III. 64 (Lyons, 1511) and Ushaw College, XVIII. F. 5. 4 (n.p. and d.).

too, to be true of the fifteenth-century binder, who, as books were produced still in comparatively small quantities, may well have been a man who was, to use a current phrase, self-employed. The individuality of many of these bindings, with their enterprising difference of design on the two covers, rather suggests this.

But when we come to the time when large numbers of books exist bearing one man's initials, the question is pertinent whether all such bindings could possibly be the work of one man. And for this we must consider not merely how many such bindings are known now, but how many once existed. Let us examine the probabilities. Duff tells us in his Sandars Lectures that of English fifteenth-century books as many as fifty-three are known only from fragments. Notoriously of many others only a single copy is recorded. Duff again tells us that of three particular editions of the New Testament of about 1533, known to have totalled some 7000 copies, not one copy now is known. These facts suggest that only a very small proportion of books, and similarly of bindings, still survive to be examined. Let us look at it from another angle. It is remarkable how often one finds only a single example of some particular tool. To take an example, we know thirty-six tools used by the Unicorn Binder on seventy books, but of these thirty-six tools, no less than eleven appear only once.[1] Is it likely that he had a tool cut and yet used it less than—let us say—ten times? If not, we must, on a moderate estimate, multiply the number of known examples by at least ten to get an idea of the number of books he originally bound. To the number of books lost, as suggested by the figures I have quoted from Duff, from wear and tear and by wholesale destruction at the Dissolution, we must add, when thinking of bindings, the enormous number rebound by collectors like Harley, and by his even less enlightened nineteenth-century librarian imitators. And what sort of numbers are extant of the work of different sixteenth-century binders? One can only of course estimate from one's own experience; I have recorded 147 examples of Godfrey, 162 of R.B. and 308 of Spierinck. If that is the number known to a single student, the number still to be discovered must obviously make the total extant enormously greater. To take the extreme case of Reynes, I have recorded 473 examples to which many others must doubtless be added. But taking merely that conservative figure and assuming that probably not more than one in ten of his original bindings survives, we get the staggering figure of 4700 bindings probably executed by him, which, if a single binder produced as many as two a week, would involve about forty-five years' continuous work, and Reynes had a publishing business as well. Clearly Reynes was not a binder, but an employer, and I suggest that, though for convenience's sake we usually call these men whose initials appear 'binders', they were often not craftsmen, but heads of firms.

The number of sets of initials used in England (apart from those on panels apparently imported from abroad) is very large. Of those to which no one has even suggested that any particular binder's name could be attached I know between sixty and seventy. Where two sets are used together on one roll they probably represent partners; when they appear on separate rolls or panels used together the simplest explanation in most cases is that a binder had picked up old signed tools and was using them indiscriminately. When there is a third initial, as in L.V.L., the natural supposition is that the last represented, as was the practice in Germany, the binder's birthplace or home town. R.H.M.I., one suspects, are nobody's initials, but stand for some motto or text.

---

[1] It is significant that, even in the sixteenth century, when books were being produced in greater numbers, of no less than ninety-seven indubitably English rolls I have found in English libraries only a single example, and of another forty-two only two examples.

Lastly, there is the question whether types of English rolls can be dated, and the problem, a purely practical and not an academic one, whether they can be classified for reference. Panel stamps, even representing the same subject, differ in the border or some other way, and are generally susceptible of recognition from a brief description. Small stamps, except the commonest types, can also often be described sufficiently well to be recognised. Not so certainly with rolls; many of them are completely indescribable. This means that, if you want to refer to a roll you have found, you must take a rubbing to the Victoria and Albert Museum and compare it physically with all that you can find like it in Weale's collection, and if you succeed in finding it (which is not as easy as it sounds) you can henceforth label it 'Weale, no. So-and-so'.

I venture to suggest, therefore, a method of classification according to the principal and subordinate features, so that each roll can be referred to by letters and a number. It is not in the least scientific, and you may regard the features I have chosen as the basis of classification as too fanciful. All I claim is that, though obviously some rolls have features common to two classes, and others are difficult to classify at all, it does in practice make the identification of a roll and reference to it far quicker and surer than is possible without some such system. And for students it is essential in making notes, for instance, of the use of one particular roll with another, to be able to use some sort of code of reference, however lacking in scientific basis.

I have divided the rolls (excluding quite indistinguishable ones of the crested and diaper type) numbering just over 400, that I know of as having been used in England, into twenty-one classes according to some noticeable feature, giving each class a two-letter reference: HM. for heads in medallions, CH. for cherub, DI. for diaper, IN. for intaglio, MW. for metalwork, FL. for floral, HE. for heraldic, AN. for animal, TC. for Tudor crested, and so on. Rolls with conventional foliage I have divided into those that are broken up into small panels, FP., and those whose design is continuous, FC., and similarly RP. and RC. for Renaissance ornament rolls. Each class I have divided up (unless it is too small to need division) according to a secondary feature, such as a capstan ornament, inscribed names, birds and so on, attaching a small letter of the alphabet to each; within each section I have numbered each roll. The basis of classification is usually simply what catches the eye first, so that a roll can be fairly easily identified and then referred to in one's notes by its letters and number. I feel I owe an apology for suggesting a system so unscholarly, but I have found it myself to save an immense amount of trouble in practice.[1]

I mentioned the question of whether these types could be dated, or, conversely, whether a binding could be dated by the type of roll used. With few exceptions I do not think much can be done in this way. The majority of types are used more or less throughout the sixteenth century. Cherub and animal rolls seem chiefly confined to the first half of the century; what I call heads in medallions with light foliage (HM. *g*) and most intaglio rolls do not appear often till about 1570. The one type of roll that is most noticeably confined to one period, the Tudor crested roll, that is, a crested roll with Tudor emblems on the top (several of which are signed), is indeed a remarkable example, for, with few exceptions, it does not appear until the seventeenth century, that is, when the Tudors had ceased to rule. I can suggest no explanation of this strange fact. One thing certainly is notable, that, except for some of the fine heraldic rolls of the end of the century, the rolls decline in artistic merit; they become narrower (these rolls requiring less skill to impress), and their designs become more stereotyped

---

[1] A classified list of rolls is given on pp. 42–58 and reproductions of them on Pls. XXXVII–LVI and LXI.

and uninspired, till they fade away into the banal and unsatisfying blind decoration of the late seventeenth and the eighteenth centuries.

I am afraid that the more one studies this subject, and attempts to disentangle the individuals who produced our earlier English bindings, and assign to them, as I have tried to do, dates, the more bewildering it all becomes. If these lectures have left you with the impression that this study involves a mass of inconsistent facts, a series of insoluble problems, a maze of paths leading in different directions and ending nowhere, that, in fact, when all has been said and a great deal done, the search for knowledge in this subject is not worth the immense labour required to reach dubious conclusions—if that is your impression from them, I am inclined to think that you are not far wrong.

APPENDIX

CLASSIFICATION OF
ROLLS AND ORNAMENTS

The following is a list of English rolls, and of ornaments that are used in the lozenge-shaped compartments into which the area within a roll-produced frame is commonly divided, grouped according to their various types. The half-ornaments, if used only in the corresponding triangular compartments, are omitted. In the small number of instances where I do not know convincing evidence that the rolls are English, though I believe them to be so, I have added '? English'. None of the ordinary crested rolls is included, nor a certain number of diaper rolls bearing quatrefoils in lozenges, because in both cases they are almost impossible to distinguish with certainty. In regard to both rolls and ornaments I have added the binder's initials, where they occur, and where the evidence seemed adequate, the name of the town to which, on present knowledge, I attribute them; most of the others were probably London tools. In some instances, however, it is clear that they moved from one town to another.

In the classification of rolls the two letters attached to each class are intended to suggest a distinctive feature of the rolls included in it, though inevitably features belonging to two classes often appear on one roll. The classes are tabulated in alphabetical order, except where the inclusion of more than one class on a single plate makes this impossible. They are as follows:

AN.	for animals. Pl. XXXVII.	
CH.	,, cherub. Pl. XXXVIII.	
DI.	,, diaper, i.e. any pattern repeating itself. Pl. XXXIX.	
FC.	,, conventional foliage, continuous. Pl. XL.	
FP.	,, conventional foliage, divided into panels. Pls. XLI, XLII.	
FR.	,, fruit. Pl. XLIII.	
FL.	,, floral design. Pls. XLIII, XLIV.	
GE.	,, German rolls used occasionally in England. Pl. LXI.	
HE.	,, heraldic design. Pls. XLV, XLVI.	
HM.	,, heads in medallions. Pls. XLVII–L.	
MW.	,, design derived from metalwork. Pl. LI.	
RC.	,, renaissance ornament, continuous. Pl. LII.	
RP.	,, renaissance ornament, divided into panels. Pl. LIII.	
CR.	,, crested rolls. (Only those that are for some reason distinctive are included.) Pl. LIV.	
TC.	,, crested rolls with Tudor emblems. Pl. LIV.	
SV.	,, rolls that are to be viewed from the side instead of the end. Pl. LV.	
IN.	,, intaglio. Pl. LV.	
SW.	,, strapwork. Pl. LVI.	
MU.	,, musical instruments. Pl. LVI.	
TP.	,, pineapples with twisted stems. Pl. LVI.	
FF.	,, rolls showing a human full figure. Pl. LVI.	

In all but a few cases these classes are subdivided into sections, to each of which a small letter (with no significance of its own) is attached, according to some secondary feature of the design.

After the class-number of each roll or ornament in the following list is given a list of the rolls, stamps and panels with which it is used, with dates. Generally those with which it is most commonly used are put first in the list. The use of the word 'Alone' only means that the roll or ornament has not been found with any tool that is here reproduced; it may have been used with crested or diaper rolls or, in the case of Godfrey and Spierinck, with stamps habitually used by them and reproduced by Gray, which are so well known that it did not seem necessary to include them.

Except where it is virtually impossible to see where a pattern begins to repeat itself (as in the case of many diaper rolls), the full length of the roll is reproduced, unless no example showing its complete length is known; in such a case the fact is mentioned in a footnote. The full width cannot always be shown, owing to the early binders' habit of running a fillet by the side of the roll and often cutting into it; this, and the fact that often even the rolls were not run straight, account for the reproductions being sometimes

AN. *a*–AN. *e* (1)   crooked. In many instances, inevitably, the rubbing reproduced is a bad one, because only a single example is known and the book is in a worn condition.

The exactness of the dates given should not be pressed; they are those in the imprint of printed books (or, if preceded by a question mark, the approximate date); in the case of blank books the dates given are those of the first entry, and as this is an unsure guide, they are preceded by a question mark, except where definite evidence exists of the date of the binding. Dates of all printed books are included, even if impossibly early to be the dates of the bindings, since omitting any would have involved, in borderline cases, an arbitrary decision as to whether a binding could or could not be as early as the date in the imprint. When only one date is given, it is not to be assumed that only one example exists; two or more may be of the same date, or there may be (and often are) others which cannot be dated.

Finally, an apology is due for the haphazard arrangement of the rolls in each class or section instead of similar ones being methodically grouped together. The reason is the purely practical one that, as, while this classification was working itself out, class-numbers had to be entered on many rubbings and in a number of my classified lists, the risk of failing to alter in all cases these references, with the consequent likelihood of serious error, seemed too great to justify any comprehensive attempt to rearrange them later.

I hope that it will be possible for other students using this book to insert in the margin of the list new tools that they find, or instances of combinations not recorded here. Some space has been allowed on Pl. LXI for pasting in rubbings of such tools. This plate is supplementary, including a few rolls either recently discovered, or such as it was not originally intended to include, on the ground of their being German.

CLASSIFIED LIST OF ROLLS

AN. ANIMALS. (PL. XXXVII)

AN. *a* EAGLE, BEE, BIRD, DOG

(1) I.R.T. London
With HM. *h* (5) [I.R.] 1495
With FP. *g* (8) [I.R.T.] 1538
With HM. *h* (6) [I.R.] 1544, ?1558
With FP. *g* (12) 1548
With FP. *d* (1) ?1558, ?1561

AN. *b* BIRD, DOG, BEE

(1) I.R. London
Alone or with ornament A (1) 1479, 1495, 1503–42
With RP. *f* (4) 1532–40
With FP. *a* (1) 1537–42
With ornament K (2) 1516
With ornament K (7) 1520, 1532
With ornament C (5) 1531
With St John the Baptist and animals in foliage panels (Hobson, *Panels*, p. 31, 1*a* and *b*) 1513
With Royal Arms and rose panels (Weale, R. 109) 1521, ?1523

(1*a*) London
Alone 1536–48

(2) London
Alone or with ornament A (1) 1482–1520
With ornament K (7) 1482, 1491, 1498, 1506
With St John the Baptist and animals in foliage panels (Hobson, *Panels*, p. 31, 1*a* and *b*) 1514

(3) London
Alone 1502–23
With CH. *a* (1) 1514, 1526–31
With DI. *c* (1) 1513–15
With SV. *a* (6) [K.L.-L.K.] 1516–28
With DI. *a* (8) 1516–28
With AN. *e* (1) [W.G.-I.G.] 1512

With FL. *a* (12) 1513
With CR. (1) 1514
With ornament F (1) 1505
With ornament A (4) 1497
With ornament D (2) 1508–14
With ornament B (1) 1511–23
With stamp 1023: 1499, 1515
With stamp 1065: 1515
With Pierre Auctorre panels (Hobson, *Panels*, p. 46, 1*a* and *b*) 1518

AN. *c* BIRD, DOG, BEE, WIVERN

(1) London
With DI. *c* (1) n.d.
With stamp 1053: 1511–21

AN. *d* BIRD, BEE, WIVERN

(1) London
Alone 1529
With DI. *b* (1) 1492, 1510
With SW. *a* (1) 1532
With AN. *n* (1) n.d.
With ornament F (2) 1528
With stamp 1037: 1494, 1510
With stamp 1062: 1510
With stamp 235: 1528
With stamp 37: 1528

AN. *e* BIRD, WIVERN

(1) W.G.-I.G. London
Alone 1514–25
With AN. *h* (1) [L.V.L.] 1503
With AN. *b* (3) 1512
With DI. *a* (9) 1515
With ornament E (1) 1506–25
With ornament B (3) ?1514
With stamps 20, 22, 24, 30: 1499–1505
With Royal Arms and Tudor emblems panels (Weale, R. 171) 1513, ?1591
With Royal Arms and rose panels signed H.I. (Weale, R. 108) ?1591

AN. *e* BIRD, WIVERN (*cont.*)

  (2) W.G.-I.G. 1520 London
      With RP. *f* (1) 1530–3
      With SW. *b* (3) ?1522
      With TP. (1) ?1522
      With ornament E (1) 1511–26
      With stamp 37: 1531–2
      With panels signed I.G. 1533
      With animals in foliage and acorn panels
        (Weale, R. 99) 1530

AN. *f* GRIFFIN, WIVERN, LION

  (1) G.G. Cambridge
      Alone 1504–34
      With DI. *a* (1) 1503–30
      With HE. *b* (1) [G.G.] 1504–21
      With HE. *b* (2) [G.G.] 1510–29
      With AN. *f* (2) [N.S.] 1515–22
      With DI. *a* (2) 1515–22
      With DI. *h* (3) 1530–3
  (2) N.S. Cambridge
      Alone 1521–33
      With FL. *a* (1) 1477, 1502–31
      With DI. *a* (2) 1493, 1502–28
      With HE. *b* (3) [N.G.] 1512–15
      With AN. *f* (1) [G.G.] 1515–22
      With HE. *b* (3*a*) [N.S.] 1518–31
      With HE. *b* (4*a*) [N.S.] 1524

AN. *g* WIVERN, LION, PORTCULLIS

  (1) London
      Alone 1487, 1496, 1506–33
      With TP. (3) 1530
      With RC. *b* (6) 1530
      With ornament B (3) 1496, 1500–5,
        1514–34
      With ornament I (2) 1504–12
      With stamps 1021, 1059: 1514
      With stamps 320, 321: n.d.

AN. *h* BIRD, WIVERN, RABBIT

  (1) L.V.L. Cambridge and ?London
      Alone 1500–4
      With AN. *e* (1) [W.G.-I.G.] 1503
      With ornament A (2) 1502–9
      With ornament B (5) 1502
      With ornament H 20 (*a*) 1497
      With stamps 20, 24: 1507
      With stamp 1051: 1506
      With stamp 1035: 1509
      With stamp 1074: 1509
      With Pieta panel signed H.I. (Hobson,
        Panels, p. 54, 9*a*) 1497
  (1*a*) Cambridge and London
      Alone 1521–2
      With RC. *a* (2) [Z.C.] 1519
      With FL. *d* (1) 1519
      With stamp 1054: 1514–27
      With stamp 1027: 1517–?1530
  (2) H.C.
      Alone 1500–13
      With ornament H 20 (*a*) 1496, 1508
      With ornament I (2) 1496, ?1502
      With ornament I (1) 1503

With ornament A (3) 1508, 1515
With ornament HS (2) 1513

AN. *i* BIRD, DOG, HARE, BEE
  (1)
      With ornament B (2) 1529
      With stamp 1041: 1529

AN. *j* BIRD, WIVERN, LION, OWL
  (1)
      Alone 1531–2
      With stamp 1025: 1531–6
      With stamp 1052: ?1522, 1531

AN. *k* WIVERN, SHEEP
  (1) London
      Alone 1513–22
      With ornament C (2) n.d.
      With stamp 1053: 1515–18

AN. *l* WIVERN
  (1) Oxford
      With ornament B (6) 1502
      With ornament H 9 (*c*) n.d.
      With stamp 153: 1505
      With stamp 322: n.d.

AN. *m* ANIMALS ENTWINED IN FOLIAGE
  (1)
      Alone 1530
      With ornament B (5) 1505
      With ornament H 16 (*b*) 1521
      With ornament F (1) n.d.
      With stamp 1061: 1496, 1505
      With stamp 1056: 1521
  (2) ?English
      Alone 1540

AN. *n* FLIES
  (1) London (probably originally French)
      With AN. *d* (1) n.d.

CH. CHERUB. (PL. XXXVIII)
  CH. *a* GROWING FLOWER AND BIRDS
    ADDORSED
  (1) London
      Alone 1521–37
      With AN. *b* (3) 1514, 1526–31
      With RC. *b* (3) 1523–39
      With SW. *b* (1) 1531
  (2)
      With FL. *b* (5) 1536
  (3) London
      With FP. *g* (10) 1557
      With DI. *d* (1) 1573
      With FC. *d* (1) 1573
  (4)
      With HM. *a* (20) 1529
      With FC. *h* (8) 1529
  (5)
      With RP. *a* (3) n.d.
      With FP. *a* (18) n.d.
  (6)
      With FP. *a* (17) 1528

CH. *b* ANIMALS

(1) London
    Alone 1542–5
    With FP. *a* (3) 1546–51
    With FC. *e* (1) 1537
(2) London
    Alone 1540–50
    With HE. *a* (1) 1540
(3) ? English
    With HM. *a* (11) 1540

CH. *c* FOLIAGE, RENAISSANCE OR NONDE-
SCRIPT ORNAMENT

(1) Oxford and probably London
    Alone 1535–41
    With HM. *h* (1) 1513–39
    With IN. (4) 1527–36
    With DI. *b* (1) 1520–6
    With HM. *h* (2) 1543
(2) W.L. London
    With HM. *h* (7) [F.P.] 1586
(3) I.B. London
    With HM. *a* (3) 1538
(4)
    Alone 1535–45
    With RP. *d* (2) 1550
(5) T.P. London
    Alone ? 1509–32
(6)
    Alone 1529
    With RP. *a* (5) 1550
(7) ? English
    With HM. *b* (4) 1533
(8) London
    With HM. *a* (22) 1509–35
    With FL. *a* (16) 1535
    With ornament B (4) 1532
(9) London
    Alone 1541, 1584
    With SW. *b* (5) 1567–70
    With FP. *g* (12) 1525
(10) London
    Alone 1533–5
    With RC. *b* (2) n.d.

DI. DIAPER. (PL. XXXIX)

DI. *a* QUATREFOILS IN LOZENGES

(1) Cambridge
    Alone 1499–1536
    With AN. *f* (1) [G.G.] 1503–30
    With HE. *b* (2) [G.G.] 1496, 1504–26
    With HE. *b* (1) [G.G.] 1519
    With ornament H 13: 1506–13
(2) Cambridge
    With HE. *b* (3) [N.G.] 1504–24
    With AN. *f* (2) [N.S.] 1493, 1502–28
    With HE. *b* (3*a*) [N.S.] 1520
    With HE. *b* (4*a*) [N.S.] 1526
(3) N.S. Cambridge
    Alone 1497, 1513
    With HE. *b* (3) [N.G.] 1512

(4) Cambridge
    Alone 1505–9
    With HE. *b* (3) [N.G.] 1499–1516
    With ornament K (3) 1505–12
    With animals in foliage panel signed
      N.C. (Hobson, *Panels*, p. 31, III),
      1511
(5) Oxford
    Alone 1545–6
    With RC. *c* (1) 1520–45
    With HM. *a* (1) [G.F.] 1528–46
    With RP. *d* (1) 1532–51
    With HM. *a* (5) 1536–43
    With MW. *a* (1) [G.K.] 1546
    With MW. *d* (11) 1546
(6)
    With ornament A (5) 1501, 1512
    With ornament D (1) 1502–16
    With ornament C (3) 1515
    With ornament HS (4) 1520
(7)
    With FP. *b* (1) 1546
    With stamp 1025: 1546
(8) London
    With AN. *b* (3) 1516–28
    With SV. *a* (6) [K.L.-L.K.] 1516,
      1528
(9) London
    With AN. *e* (1) [W.G.-I.G.] 1515
    With ornament HS (2) 1514–24
(10) London
    Alone 1551, 1574
    With HM. *h* (8) 1527–35, 1553–80
    With HM. *b* (3) 1527, 1531, 1558–80
(11) London
    With HM. *g* (7) 1551–75
(12) London
    With HM. *g* (6) 1573
(13) London
    Alone ? 1593
    With CR. (3) [T.M.] 1609

DI. *b* FLEURS-DE-LIS IN LOZENGES. (Fault:
every sixth upside down)

(1) London
    With HM. *h* (1) 1529–40
    With AN. *d* (1) 1492, 1510
    With CH. *c* (1) 1520–6
    With ornament A (2) 1501
    With ornament H 20 (*a*) 1524
    With stamp 336: 1504
    With stamp 1037: 1510
    With stamp 1062: 1510
    With stamp 1031: 1524
    With Royal Arms panel signed F.L.
      (Hobson, *Panels*, p. 41, VI), 1523–8
    With Tudor emblems panel signed T.L.
      (Hobson, *Panels*, p. 41, VII),
      1528
    With Flagellation panel (Hobson, *Panels*,
      p. 54, no. 7) 1523
    With animals in circles panel (Hobson,
      Panels, p. 54, under no. 7) 1523

DI. *c* Sexfoils between undulating
  lines
(1) London
    With AN. *b* (3) 1513–15
    With AN. *c* (1) n.d.
    With ornament B (1) 1488, 1504
    With stamp 1042: 1500
    With stamp 1043: 1500
    With stamp 1044: 1500

DI. *d* Fleurs-de-lis and flowers in
  lozenges alternating
(1) London. (Fault: two fleurs-de-lis together)
    With CH. *a* (3) 1573
    With FC. *d* (1) 1573
(2) London. (Fault: two flowers together)
    With HE. *g* (3) [H.R.] 1483
    With HM. *h* (21) ?1591–?1596
    With FP. *f* (5) ?1591–?1596

DI. *e* Fleurs-de-lis and flowers in ovals
  alternating
(1)
    Alone 1487, 1525, 1538, 1543
(2) ?London
    Alone 1547
    With RP. *a* (5) 1555
(3) L.W. Norwich
    Alone 1556
    With FP. *g* (13) ?1541–63

DI. *f* Cinqfoils in ovals
(1) London
    With FL. *a* (8) 1554

DI. *g* Roses in lozenges
(1) London
    With HM. *a* (6) 1550–63

DI. *h* Nondescript
(1) G.C.
    Alone 1565–9
    With FC. *h* (5) 1500, 1558–70
(2) London
    Alone 1623
(3) Cambridge and London
    Alone 1521–37
    With AN. *f* (1) [G.G.] 1530–3
    With HE. *b* (2) [G.G.] 1533–8
    With FP. *g* (4) ?1524, 1529
    With CR. (2) 1530
(4) London
    With HE. *g* (2) [H.R.] 1517

FC. FOLIAGE: CONTINUOUS.
(PLS. XL, LXI)
FC. *a* Geometrical
(1)
    With FC. *h* (5) 1546–63
(2) London
    With RP. *b* (1) 1561
    With HM. *e* (1) 1573
    With HM. *g* (4) 1573

(3) London
    With HM. *b* (3) [F.D.] 1596
    With RC. *a* (1) 1596
    With HE. *d* (1) 1596

FC. *b* Spiral
(1)
    Alone 1506
(2) London
    Alone 1521–3
    With ornament K (1) 1513–23
    With stamp 1053: 1519–23
    With stamp 1042: 1519
    With stamp 1044: 1519

FC. *c* Entwined twig
(1) Oxford
    Alone 1523–38, 1554–7
    With HM. *d* (1) 1537–41
    With MW. *b* (1) 1533
    With RP. *d* (1) 1533, 1550
    With MW. *d* (9) 1537
(2)
    With HM. *d* (2) 1516
    With ornament B (7) 1516
(3) London
    Alone n.d.

FC. *d* Flowers in circles
(1) London
    With CH. *a* (3) 1573
    With DI. *d* (1) 1573

FC. *e* Head and skull
(1) London
    Alone 1540–5
    With RC. *b* (3) [R.W.] 1534
    With CH. *b* (1) 1537

FC. *f* Helmeted head
(1) London
    Alone 1528
    With HE. *b* (5) ?1508

FC. *g* Vase
(1) London
    With HM. *a* (3) 1557
    With HE. *a* (10) 1557
    With HM. *a* (15) 1566
    With HM. *a* (4) ?1600
    With HM. *e* (2) ?1600
    With HE. *a* (1) ?1600
    With SW. *b* (2) ?1600
(2)
    Alone 1542

FC. *h* Nondescript
(1) London
    Alone 1548
    With HM. *b* (1) 1498
(2) R.W.
    Alone 1526–45
    With FC. *h* (3) [R.W.] 1529

FC. *h* NONDESCRIPT (*cont.*)

(3) R.W.
With FC. *h* (2) 1529

(4)
With HM. *a* (17) 1514

(5)
Alone ? 1563
With FC. *a* (1) 1546–63
With DI. *h* (1) [G.C.] 1500, 1558–70

(6) ? London
With HM. *g* (10) 1571
With GE (2) 1519, 1547, 1553, 1564–80
With stamps 1066–72: 1534

(7) R.W.[1]
Alone 1544

(8)
With HM. *a* (20) 1529
With CH. *a* (4) 1529

(9)
With HM. *f* (1) 1530–7

(10) R.C.[1] London
With TP. (1) n.d.

(11) London
Alone 1556
With HM. *h* (10) [R.B.] 1551–7

(12) Norwich
With IN. (5) 1548

FP. FOLIAGE: PANELLED. (PLS. XLI, XLII, LXI)

FP. *a* GROWING FLOWER

(1) London
Alone 1539–44
With AN. *b* (1) [I.R.] 1537–42

(2) London
Alone 1537–40
With HM. *a* (12) 1540

(3) London
Alone 1536–78
With RC. *b* (4) [N.E.] 1537–45
With CH. *b* (1) 1546–51

(4) London
Alone 1540–1
With RC. *b* (3) [R.W.] 1520–41
With HE. *c* (1) 1543

(5) London
Alone ? 1544
With HE. *a* (6) [I.S.] ? 1614

(6) F.I. London
Alone 1535–49
With HM. *a* (7) 1531–8
With FL. *a* (5) 1532–9
With RP. *a* (5) 1538–51
With FL. *b* (1) ? 1520

(7) London
With SW. *b* (3) ? 1516–39
With FP. *a* (20) 1535

(8)
Alone 1525–50
With HM. *a* (17) 1495, 1519–51

(9) London
With HM. *a* (4) ? 1516
With TP. (2) ? 1516

(10) ? Oxford
With MW. *d* (11) 1551

(11)
Alone n.d.

(12) ? English
With HM. *a* (16) 1531

(13) London
Alone 1548
With HE. *g* (2) [H.R.] 1547

(14)
With HM. *e* (5) 1542

(15) I.W. Oxford
Alone 1514, 1572, 1598–1622
With HM. *h* (4) 1606–17

(16)
With HM. *c* (7) 1537

(17)
Alone 1545
With CH. *a* (6) 1528

(18)
With CH. *a* (5) n.d.
With RP. *a* (3) n.d.

(19)
Alone 1534

(20) London
With FP. *a* (7) 1535
With SW. *b* (3) 1535

(21) ? English
With HM. *a* (19) 1528

(22)
With HM. *a* (18) 1547

(23)
Alone 1544

(24)[2]
Alone 1563

FP. *b* BUDS

(1)
With DI. *a* (7) 1546
With stamp 1025: 1546

(2)
Alone 1529–42

(3) London
With HM. *c* (6) 1533
With TC. *a* (5) ? 1562, 1581, ? 1615

(4)
Alone 1548

(5)
Alone 1540

(6) H.R. London
With FP. *e* (1) ? 1543

FP. *c* LARGE HEADS

(1) London
With HM. *c* (1) 1502, 1538
With HM. *b* (1) 1537

[1] The roll as reproduced is not complete.
[2] The only known example, in the Folger Library, is too worn to reproduce. This roll and FP. *a* (23) were brought to my notice by Mr J. G. McManaway.

ENGLISH BLIND-STAMPED BINDINGS

FP. *d* Square cartouche

(1) London
    Alone 1534–46
    With HM. *h* (6) [I.R.] ?1558
    With AN. *a* (1) [I.R.T.] ?1558–?1561
    With FP. *f* (8) 1530

FP. *e* Half-wheels

(1)
    Alone ?1542
    With FP. *b* (6) ?1543

FP. *f* Birds

(1) London
    Alone 1512, 1562
(2) London
    With HM. *c* (4) 1545
(3) London
    With HM. *a* (3) ?1563
    With HE. *a* (10) ?1563
(4) D.V. ?Oxford
    With ornament H 9 (*b*) ?1493
    With stamp 146: 1501
    With stamp 1032: 1512
    With stamp 1049: 1512
    With stamp 37: n.d.
(5) London
    Alone 1547
    With HE. *g* (3) [H.R.] 1533, 1562–80
    With HM. *h* (21) ?1591–?1596
    With DI. *d* (2) ?1591–?1596
    With MW. *d* (2) 1576
    With FP. *g* (11) 1576
(6) London
    Alone 1525–52
    With HM. *e* (1) ?1557
    With FL. *a* (6) ?1557
(7) London
    Alone 1551, 1620
    With HM. *a* (23) 1507, 1597–1605
(8) London
    With FP. *d* (1) 1530
(9) London
    Alone 1553–?1560
(10) London
    Alone 1556
    With HE. *a* (10) 1557
(11) London
    With HM. *h* (11) 1567

FP. *g* Nondescript

(1) ?London
    Alone 1543–4
    With FL. *a* (1) 1536
    With FP. *g* (4) 1536
    With stamp 94: 1536
(2) London
    Alone 1535–51
    With RC. *b* (6) 1530
    With TP. (3) 1534
    With SW. *b* (6) 1534
    With ornament A (6) 1517

(3) London
    Alone ?1567
    With SV. *a* (3) 1568
(4) London
    With DI. *h* (3) ?1524–9
    With FL. *a* (1) 1536
    With FP. *g* (1) 1536
    With HE. *b* (2) [G.G.] 1537
    With stamp 94: 1536
(5) R.P. London
    Alone n.d.
    With MW. *d* (1) [R.P.] 1512, 1551–70
    With HM. *h* (20) 1553
    With HE. *b* (6) 1569–70
    With HM. *a* (14) 1569
(6) and (6*a*) R.W. Oxford
    Alone 1518, 1560, 1596–1620
    With MW. *d* (11) 1610
(7) I.A. Oxford
    Alone 1555, 1600–19
    With RP. *d* (1) 1598–1612
    With MW. *d* (11) 1613
(8) I.R.T. London
    Alone 1486, 1500, 1552, ?1603, ?1628
    With AN. *a* (1) [I.R.T.] 1538
(9) Oxford
    Alone 1590–1605
    With RP. *d* (1) 1544–55
    With MW. *a* (1) [G.K.] 1513–54, 1569–
      1605
    With MW. *b* (1) 1544
    With RC. *b* (1) 1564
(10) London
    Alone 1530–48
    With CH. *a* (3) 1557
(11) London
    With HE. *g* (3) [H.R.] 1576
    With FP. *f* (5) 1576
    With MW. *d* (2) 1576
(12) London
    With CH. *c* (9) 1525
    With AN. *a* (1) [I.R.T.] 1548
    With SV. *a* (3) 1556
(13) Norwich and London
    Alone 1553–65, 1603
    With DI. *e* (3) [L.W.] ?1541–63
    With IN. (5) ?1561
    With TC. *b* (3) [H.C.] 1634
(14)
    With HM. *h* (3) 1552
(15) N.D.-B.B. London
    Alone 1538

FR. FRUIT. (PL. XLIII)
FR.

(1) Oxford
    With ornament H. 20 (*a*) 1502
    With stamp 146: 1495–1501
    With stamp 154: 1495–1502
    With stamp 153: 1501–2
    With stamp 149: 1502
    With stamp 30: 1502

FP. *d* (1)–FR. (1)

47

FR. (1)–FL. *b* (8)

FR. (*cont.*)

    With stamp 1029: 1502
    With stamp 156: n.d.

(2)

    Alone 1482, 1556–76

(3)

    With ornament D (2) 1504
    With ornament H 16 (*b*) 1510
    With ornament H 16 (*c*) ? 1512
    With stamp 1026: 1504
    With stamp 1047: ? 1512

FL. FLORAL. (PLS. XLIII, XLIV)

FL. *a* CONTINUOUS STEM

(1) Cambridge and London
    Alone 1512–43
    With AN. *f* (2) [N.S.] 1477, 1502–31
    With HM. *h* (28) 1529–45
    With HE. *b* (4) [I.S.] 1511
    With HE. *b* (4*a*) 1524
    With HE. *b* (3*a*) [N.S.] 1531
    With HM. *h* (29) 1536
    With FP. *g* (1) 1536
    With FP. *g* (4) 1536
    With stamp 94: 1536

(2) London
    Alone 1495, 1530, ? 1540
    With Tudor emblems panel 1565

(3) Oxford
    Alone 1500–21
    With ornament H. 9 (*a*) 1500–17
    With ornament H. 9 (*c*) 1513–15

(4) London
    Alone 1508–28
    With RC. *e* (1) 1519–30
    With MU. (2) 1521–5
    With ornament B (5) 1492, 1494, 1508–27
    With ornament A (4) 1521
    With ornament A (9) 1522
    With ornament C (1) 1526
    With stamp 1060: n.d.
    With stamp 1040: 1522
    With stamp 1063: 1522
    With stamp 1030: 1523
    With stamp 1039: 1523–5
    With stamp 1025: 1497
    With heads in medallions panels signed G.P. (Hobson, *Panels*, p. 49, 11 *a* and *b*) 1529
    With Royal Arms and rose panels (Weale, R. 109) 1502–12

(5) London
    With FP. *a* (6) [F.I.] 1532–9
    With RP. *a* (1) 1533
    With HM. *a* (7) 1534
    With HM. *f* (1) 1536
    With RP. *a* (5) 1538

(6) London
    Alone 1519–28
    With RP. *f* (2) 1498

    With HM. *e* (1) ? 1557
    With FP. *f* (6) ? 1557
    With four-compartment panel signed S.G. (2nd state) (Weale, R. 158) 1530
    With flowers and acorns panel (Weale, R. 159) 1530

(7) ? Oxford and London
    Alone 1536
    With RP. *c* (1) 1559–64
    With HM. *h* (3) 1562–6
    With RC. *b* (3) [R.W.] 1539
    With HM. *a* (3) 1560
    With RC. *b* (1) 1562

(8) London[1]
    Alone ? 1545–6
    With DI. *f* (1) 1554

(9)
    Alone 1531–41

(10)
    Alone 1544

(11) London
    With RC. *b* (4) [N.E.] 1571–3
    With HE. *a* (4) [I.H.] ? 1575, ? 1593

(12) London
    With AN. *b* (3) 1513
    With stamp 1023: 1513

(13)[1]
    With ornament HS (2) 1519, 1528

(14)[1]
    Alone 1534

(15) London
    With stamp 1057: 1521
    With Royal Arms and rose panels (Weale, R. 109) 1521

(16) London
    With HM. *a* (22) 1535
    With CH. *c* (8) 1535

FL. *b* SPIRAL STEM

(1) London
    Alone 1535, 1557
    With HM. *a* (7) 1532–6
    With FP. *a* (6) [F.I.] ? 1520

(2) ? English[1]
    Alone 1521

(3) ? London
    With ornament B (4) 1522

(4) London
    Alone 1519
    With HM. *h* (6) [I.R.] 1602
    With HE. *i* (1) 1602

(5)
    With CH. *a* (2) 1536

(6)
    With ornament HS (2) 1526

(7) London
    With HE. *c* (1) 1583

(8) ? London
    With ornament B (10) 1508
    With ornament C (5) 1516

---

[1] The roll as reproduced is not complete.

FL. *c* FLOWER SPRAY REPEATED
(1) Oxford
  Alone 1578
  With MW. *d* (10) 1572–? 1605
  With RP. *c* (1) 1598–1630
  With HM. *d* (1) 1545
(2) Oxford
  Alone 1556–7
(3) Oxford[1]
  Alone n.d.

FL. *d* SCROLL-WORK
(1) ? Cambridge
  With RC. *a* (2) [Z.C.] 1519–29
  With AN. *h* (1 *a*) 1519
  With SW. *b* (3) 1529
(2) London
  With HE. *g* (4) 1531, 1552–6

FL. *e* FLOWERS IN CIRCLES
(1)
  Alone 1521
(2)
  Alone n.d.

FL. *f* NONDESCRIPT
(1) ? Oxford[2]
  With HM. *h* (1) 1533
  With ornament J (3) n.d.

GE. GERMAN. (PL. LXI)
(1) Iuno, Venus, Paris, Pallas, 1539. (Haebler, vol. II, p. 13, no. 3) London
  With TC. *a* (3) 1571–84
  With TC. *a* (7) 1609
(2) Lucret, Suavita, Pruden.[3] ? Cambridge
  With FC. *h* (6) 1519, 1547, 1553, 1564, 1571–80
  With HM. *g* (10) 1519, 1547, 1553, 1564, 1571–80
(3) Lucreti, Suavitas, Prudent. ? Cambridge
  With HE. *g* (4) 1529, 1551
  With HM. *g* (8) 1539, 1541, 1592
  With GE. (4) 1498, 1504, 1523–9, 1542, 1557–62, 1575–84
  With IN. (10) 1575
(4) Lucreti, Suavitas, Pruden, Iustit. ? Cambridge
  Alone 1581–9
  With HE. *g* (4) ? 1573
  With GE. (3) 1498, 1504, 1523–9, 1542, 1557–62, 1575–84

(5) Fides, Spes, Caritas, Fortit. London
  With HM. *a* (21) 1572
  With HM. *f* (2) 1572
  With HM. *h* (11) 1572
(6) Fides-Nonco, Esperatia, Caritas. London
  With MW. *d* (13) 1572
(7)[4] Heads in oval medallions. London
  With HE. *a* (4) ? 1628
  With HE. *a* (8) ? 1628
(8) Arms of Saxony (Weale, R. 719). London
  Alone, 1527
(9) Gedion, Josua. London
  With FP. *a* (9) 1582

HE. HERALDIC. (PLS. XLV, XLVI)
HE. *a* ROYAL ARMS
(1) London
  Alone 1521–59
  With RP. *a* (1) 1544–54
  With RP. *f* (2) 1540
  With CH. *b* (2) 1540
  With HM. *e* (2) ? 1583, ? 1600
  With HE. *b* (5) ? 1585, 1592
  With SV. *b* (2) ? 1585
  With RC. *b* (4) [N.E.] ? 1586
  With HM. *a* (4) ? 1600
  With FC. *g* (1) ? 1600
  With SW. *b* (2) ? 1600
  With ornament C (5) 1521–30
(2) London
  Alone 1588, ? 1614
(3) London
  Alone 1586
  With HM. *a* (3) 1580–5
(4) I.H. London
  Alone 1588–? 1606
  With FL. *a* (11) ? 1575–? 1593
  With HE. *a* (8) [G.D.] ? 1628
  With GE. (7) ? 1628
(5) E.P. London
  Alone ? 1622–1636
(6) I.S. London
  With HM. *a* (7) 1590
  With FP. *a* (5) ? 1614
(7) L.P. London
  Alone ? 1594
(8) G.D. London
  With HE. *a* (4) [I.H.] ? 1628
  With GE. (7) ? 1628
(9) London
  Alone 1583

---

[1] No impression has been found good enough for reproduction.

[2] The roll as reproduced is not complete.

[3] GE. (2), (3) and (4) were recently brought to my notice by Mr N. Ker. These books all belonged to John Betts, they all have hatching on the back and on the edges at head and tail (see *S.S.L.B.* pp. 20–1), and some are linked through a flowers-in-vase stamp with bindings that bear a large ornate stamp whose centre is sometimes filled with the Cambridge University arms. The binder was evidently a German immigrant, as on Emmanuel College, Cambridge, 318. 3. 10–16, though the endpapers are fragments of an English MS., he not only put the clasps on the lower cover, but retained the entirely non-English feature of the owner's initials at the top, and the date at the bottom, which is a characteristic of German bindings.

[4] The only known example, in the Folger Library, is too worn to reproduce.

HE. *a* ROYAL ARMS (*cont.*)
    With HM. *h* (6) [I.R.] 1537, 1598
    With HE. *i* (1) 1537, 1598
(10) London
    Alone 1516, 1537–9, 1556–63, 1586
    With HM. *a* (3) 1543–64
    With FP. *f* (10) 1557
    With FC. *g* (1) 1557
    With FP. *f* (3) ?1563
(11) I.R. London
    Alone 1623

HE. *b* GATEWAY, FLEUR-DE-LIS, POME-
    GRANATE, ROSE
(1) G.G. Cambridge
    With AN. *f* (1) [G.G.] 1504–21
    With DI. *a* (1) 1519
(2) G.G. Cambridge
    Alone 1501–36
    With DI. *a* (1) 1496, 1504–26
    With AN. *f* (1) [G.G.] 1510–29
    With DI. *h* (3) 1533–8
    With FP. *g* (4) 1537
    With ornament H 13 (*a*) 1505–12
    With ornament I (2) 1511
(3) N.G. Cambridge
    Alone 1493, 1502–15
    With DI. *a* (4) 1499–1516
    With AN. *f* (2) [N.S.] 1512–15
    With DI. *a* (2) 1504–24
    With DI. *a* (3) [N.S.] 1512
    With animals in foliage panel signed
      N.C. (Hobson, *Panels*, p. 31, III) n.d.
(3*a*) N.S. Cambridge
    With AN. *f* (2) [N.S.] 1518–31
    With DI. *a* (2) 1520
    With FL. *a* (1) 1531
    With stamp 94: 1528
(4) I.S. Cambridge
    Alone 1508–20
    With FL. *a* (1) 1511
    With SV. *a* (1) 1522
    With SS. John the Baptist and Roch
      panels (Hobson, *Panels*, p. 48, 1*a*
      and *b*), 1516
(4*a*) N.S. Cambridge
    With AN. *f* (2) [N.S.] 1524
    With FL. *a* (1) 1524
    With DI. *a* (2) 1526
(5) London
    Alone 1487, 1571–93
    With HM. *h* (10) [R.B.] 1541–61, 1581
    With MW. *d* (12) 1555–70
    With MW. *c* (1) 1559–71
    With HE. *k* (1) 1556, 1570–86
    With SV. *a* (8) 1580–4
    With HM. *h* (20) ?1585–88
    With FC. *f* (1) ?1508
    With HE. *a* (1) ?1585, 1592
    With SV. *b* (2) ?1585
    With HM. *b* (5) ?1585
    With TC. *a* (7) n.d.

(6) London
    Alone 1586
    With HM. *a* (14) 1525, 1569–72
    With FP. *g* (5) [R.P.] 1569–70
    With MW. *d* (1) [R.P.] 1570
    With HM. *a* (9) 1525, 1573

HE. *c* GATEWAY, ROSE, FLEUR-DE-LIS
(1) London
    Alone 1516, 1545, 1574, ?1583
    With FP. *a* (4) 1543
    With RC. *b* (6) 1544
    With FL. *b* (7) 1583

HE. *d* GATEWAY, PORTCULLIS, LION, ROSE,
FLEUR-DE-LIS, POMEGRANATE, ENTWINED
IN FOLIAGE
(1) London
    Alone 1512–18
    With HM. *d* (5) ?1520
    With HM. *b* (3) [F.D.] 1596
    With FC. *a* (3) 1596
    With RC. *a* (1) 1596
    With ornament A (2) 1505–27
    With Royal Arms and rose panels
      signed H.I. (Weale, R. 107, R. 108)
      1520
(2) London
    Alone 1515–17
    With Royal Arms and Tudor emblems
      panels (*S.S.L.B.* p. 95) 1518

HE. *e* PORTCULLIS, CROWN, ROSE, ANCHOR,
POMEGRANATE, ALTERNATING WITH
FLEUR-DE-LIS
(1) G.L.
    Alone ?1648

HE. *f* PORTCULLIS, HARP, FLEUR-DE-LIS,
ROSE, ALTERNATING WITH LION,
PEGASUS, DOLPHIN, EAGLE IN CIRCLES
(1) E.R. London
    With HM. *h* (13) 1521

HE. *g* PORTCULLIS, ROSE, FLEUR-DE-LIS
(1) H.R. London
    Alone 1551, 1654
(2) H.R. London
    Alone 1545–72
    With SW. *b* (5) 1553–74
    With MW. *d* (8) 1550–5
    With DI. *h* (4) 1517
    With FP. *a* (13) 1547
(3) H.R. London
    Alone 1518, 1535–46, ?1561, ?1566,
      1580–?1584, 1606, ?1631, ?1635
    With FP. *f* (5) 1533, 1562–80
    With FP. *g* (11) 1576
    With MW. *d* (2) 1576
    With DI. *d* (2) 1483
(4) London and Cambridge
    Alone 1504, 1521–67, 1584
    With FL. *d* (2) 1531, 1552–6
    With RP. *a* (2) 1550

HE. *g* PORTCULLIS, ETC. (*cont.*)
  With IN. (10) 1550
  With HM. *g* (8) 1590
  With GE. (3) 1573
  With GE. (4) 1573
  With HM. *g* (2) 1574
  With ornament C (4) 1526–34

HE. *h* PORTCULLIS, POMEGRANATE, ROSE
  (1) London
    Alone 1516–41
    With ornament C (4) 1511–35
    With ornament C (5) 1521–30
    With SS. John the Baptist and Roch
      panels (Hobson, *Panels*, p. 48, 1*a*
      and *b*), ?1530

HE. *i* NON-ROYAL ARMS
  (1) R.F. London
    Alone 1587
    With HM. *h* (6) [I.R.] 1537, 1598–1602
    With HE. *a* (9) 1537, 1598
    With FL. *b* (4) 1602

HE. *j* PORTCULLIS
  (1) London
    Alone 1509, 1557, 1562
  (2) London[1]
    With small panel of head 1024: 1527

HE. *k* GOLDEN FLEECE AND FALCON
  (1) London
    Alone 1545–67, 1581–98
    With HM. *h* (10) [R.B.] 1527, ?1538,
      1556–68
    With MW. *a* (4) 1550–62
    With HE. *b* (5) 1556, 1570–86
    With HM. *b* (5) ?1549, ?1585
    With MW. *d* (8) 1550
    With HM. *h* (20) ?1585
    With HM. *a* (3) 1589

HE. *l* TUDOR EMBLEMS IN CIRCLES
    BETWEEN LOVE-KNOTS
  (1) London
    With CR. (4) [E.B.] 1610
    With TC. *a* (3) 1613–23
  (2) London
    With TC. *b* (1) [T.E.] 1613
  (3) E.R. London
    With TC. *a* (6) 1592

HM. HEADS IN MEDALLIONS.
(PLS. XLVII–L, LXI)
HM. *a* CAPSTAN
  (1) G.F. Oxford
    Alone 1541–4
    With RC. *c* (1) 1528–44
    With DI. *a* (5) 1528–46
    With HM. *a* (5) 1538–43
    With HM. *a* (1) 1546

  (2) London
    Alone 1518–33
  (3) London
    Alone 1546–88
    With HE. *a* (10) 1543–64
    With HM. *a* (6) 1548–80
    With HE. *a* (3) 1580–5
    With CH. *c* (3) [I.B.] 1538
    With FC. *g* (1) 1557
    With FL. *a* (7) 1560
    With FP. *f* (3) ?1563
    With HE. *k* (1) 1589
  (4) London
    With SW. *b* (2) 1523–31, 1552–9, ?1600
    With HM. *a* (7) 1531–59
    With RP. *b* (1) 1504
    With TP. (2) ?1516
    With FP. *a* (9) ?1516
    With HM. *e* (2) ?1600
    With HE. *a* (1) ?1600
    With FC. *g* (1) ?1600
  (5) Oxford
    With DI. *a* (5) 1536–43
    With HM. *a* (1) 1538–43
    With RC. *e* (1) 1537
    With ornament C (1) 1532
    With ornament J (3) ?1537
  (6) London
    Alone 1530–42, 1551–80
    With HM. *a* (3) 1548–80
    With MW. *d* (12) 1527, 1562
    With MW. *d* (6) 1529, 1546, 1556
    With DI. *g* (1) 1550, 1563
    With SW. *b* (5) 1568, 1573
  (7) London
    Alone 1523–57
    With SW. *b* (2) 1523–31, 1559
    With HM. *a* (4) 1531–3, 1550–9
    With FP. *a* (6) [F.I.] 1531–8
    With FL. *b* (1) 1532–6
    With RC. *b* (6) 1536–8
    With RP. *a* (5) 1523, 1531
    With FL. *a* (5) 1534
    With HE. *a* (6) [I.S.] 1590
  (8)
    With ornament J (2) 1521
  (9)
    With HE. *b* (6) 1525, 1573
  (10) London and Oxford
    Alone 1647
    With RP. *a* (4) 1532–49
  (11) ?English
    With CH. *b* (3) 1540
  (12) London
    Alone 1530–8
    With FP. *a* (2) 1540
  (13) London
    Alone 1530–69
    With MW. *d* (14) 1531
  (14) London
    With HE. *b* (6) 1525, 1569–72
    With FP. *g* (5) 1569

HE. *g* (4)–
HM. *a* (14)

---

[1] The roll as reproduced is not complete.

HM. *a* (15)–
HM. *e* (5)

HM. *a* CAPSTAN (*cont.*)

(15) London
Alone 1563–9
With FC. *g* (1) 1566
(16) ?English
Alone 1527
With FP. *a* (12) 1531
(17)
Alone 1490, 1523–45, 1587
With FP. *a* (8) 1495, 1519, 1530–51
With HM. *h* (24) 1551–64
With FC. *h* (4) 1514
(18)
With FP. *a* (22) 1547
(19) ?English
With FP. *a* (21) 1528
(20)
With CH. *a* (4) 1529
With FC. *h* (8) 1529
(21) London
Alone 1561
With HM. *f* (2) 1572
With HM. *h* (11) 1572
With GE. (5) 1572
(22) London
With CH. *c* (8) 1509, 1531–5
With FL. *a* (16) 1535
With ornament B (4) 1532
(23) London
Alone 1570
With FP. *f* (7) 1507, 1597, ?1605

HM. *b* NAMES

(1) [HERCULES, VENUS] London
Alone 1531, 1549
With HM. *c* (1) ?1526–37
With FC. *h* (1) 1498
With FP. *c* (1) 1537
(2) [HERCULIS, VEHA] London[1]
Alone 1527, 1585
(3) F.D. London
Alone 1554–76
With DI. *a* (10) 1527–31, 1558–80
With SW. *b* (5) 1570–4
With HM. *h* (8) 1580
With HE. *d* (1) 1596
With FC. *a* (3) 1596
With RC. *a* (1) 1596
(4) ?English
With CH. *c* (7) 1533
(5) London
Alone 1507
With HE. *k* (1) ?1549, ?1585
With HE. *b* (5) ?1585
With HM. *h* (20) ?1585

HM. *c* BIRDS

(1) London
Alone 1538, 1556
With HM. *b* (1) ?1526–1537

With FP. *c* (1) 1502, 1538
With RC. *b* (2) n.d.
(2)
Alone 1509, 1536
(3) London
Alone 1549
(4) London
Alone 1544
With HM. *h* (5) [I.R.] 1477, 1579–83
With FP. *f* (2) 1545
(5)[2]
Alone 1539, 1553
(6)
With FP. *b* (3) 1533
With RC. *b* (6) n.d.
(7)
With FP. *a* (16) 1537

HM. *d* BEARDED HEADS

(1) Oxford and ?London
Alone 1506, 1530–43, 1573
With FC. *c* (1) 1537–41
With MW. *d* (9) 1537
With FL. *c* (1) 1545
With MW. *a* (1) 1559
With MW. *b* (1) 1527, 1561
(2)
With FC. *c* (2) 1516
With ornament B (2) 1518–28
With ornament B (7) 1516
(3) ?English
Alone 1530
(4)
Alone 1522–37
With CR. (2) 1528
(5)
With HE. *d* (1) ?1520

HM. *e* CHERUB

(1) London
With HM. *g* (4) 1573–87
With FP. *f* (6) ?1557
With FL. *a* (6) ?1557
With FC. *a* (2) 1573
(2) London
With HE. *a* (1) ?1583, ?1600
With HM. *a* (4) ?1600
With SW. *b* (2) ?1600
With FC. *g* (1) ?1600
(3) ?English
Alone 1532–41
With HM. *e* (4) 1531
(4) ?English
With HM. *e* (3) 1531
(5)
Alone 1523–5
With FP. *a* (14) 1542
With HM. *h* (31) 1546

[1] Mr N. Ker tells me that Eton College accounts show that the binder of these books was named Williamson.
[2] The roll as reproduced is probably not complete.

HM. *f* FLOWERS IN MEDALLIONS

(1)

> With RC. *b* (2) 1515–25
> With FC. *h* (9) 1530–7
> With FL. *a* (5) 1536

(2) London

> Alone 1535–54
> With HM. *a* (21) 1572
> With HM. *h* (11) 1572
> With GE. (5) 1572

HM. *g* LIGHT FOLIAGE

(1) London

> Alone 1572–4
> With TC. *a* (2) 1506, 1577–84, 1599–
> 1627, ? 1647

(2) ? Cambridge[1]

> Alone 1584
> With HE. *g* (4) 1574
> With IN. (9) 1577

(3) London

> Alone 1571

(4) London

> Alone 1583
> With HM. *e* (1) 1573–87
> With FC. *a* (2) 1573

(5) London

> With HM. *h* (17) 1524

(6) London

> With DI. *a* (12) 1573

(7) London

> With DI. *a* (11) 1551-74

(8) London

> Alone 1570–81
> With HE. *g* (4) 1590
> With GE. (3) 1539, 1541, 1592

(9)

> Alone 1573, ? 1593

(10)[2]

> With FC. *h* (6) 1571
> With GE. (2) 1519, 1547, 1553, 1564–80

HM. *h* FOLIAGE, RENAISSANCE, OR NON-
DESCRIPT ORNAMENT

(1) Oxford, and probably London

> Alone 1504–41
> With DI. *b* (1) 1529–40
> With IN. (4) 1534–41
> With CH. *c* (1) 1513–39
> With FL. *f* (1) 1533
> With HM. *h* (15) 1534
> With ornament C (1) 1536

(2) Oxford

> Alone 1531–44
> With MW. *d* (10) 1586–? 1605
> With CH. *c* (1) 1540–3
> With IN. (11) 1603

(3) ? Oxford

> Alone 1560
> With FL. *a* (7) 1562–6

With RP. *c* (1) 1562–4
With FP. *g* (14) 1552

(4) Oxford

> Alone 1602–? 1605
> With MW. *d* (9) 1532–50
> With IN. (3) 1532
> With FP. *a* (15) [I.W.] 1606–17
> With MW. *a* (1) [G.K.] 1473
> With RP. *d* (1) ? 1605

(5) I.R. London

> Alone 1551, 1615
> With HM. *c* (4) 1477, 1579, 1583
> With AN. *a* (1) [I.R.T.] ? 1495
> With HM. *h* (6) [I.R.] 1549
> With IN. (8) 1568

(6) I.R. London

> Alone 1550, 1574
> With HE. *a* (9) 1537, 1598
> With HE. *i* (1) 1537, 1598, 1602
> With AN. *a* (1) [I.R.T.] 1544, ? 1558
> With HM. *h* (5) [I.R.] 1549
> With FP. *d* (1) ? 1558
> With FL. *b* (4) 1602

(7) F.P. London

> With MW. *d* (7) 1549
> With CH. *c* (2) 1586

(8) London

> Alone 1578
> With DI. *a* (10) 1527–35, 1553–80
> With HM. *b* (3) [F.D.] 1580

(9)

> Alone 1552, 1572

(10) R.B. London

> Alone 1537–69
> With HE. *b* (5) 1541–61, 1581
> With HE. *k* (1) 1527, ? 1538, 1556–68
> With MW. *d* (12) 1550–70
> With FC. *h* (11) 1551-7
> With IN. (1) 1573
> With IN. (6) 1578

(11) London

> Alone 1555–60
> With RP. *b* (1) 1549
> With FP. *f* (11) 1567
> With HM. *a* (21) 1572
> With HM. *f* (2) 1572
> With HM. *h* (12) 1573
> With GE. (5) 1572

(12) London

> With HM. *h* (11) 1573

(13) London

> With HE. *f* (1) [E.R.] 1521

(14) R.D. ? London

> Alone ? 1583
> With TC. *a* (6) 1499
> With IN. (9) 1544

(15)

> Alone 1527
> With RC. *b* (5) 1530–1
> With RC. *b* (6) 1531

HM. *f* (1)–
HM. *h* (15)

[1] Mr J. C. T. Oates tells me that this roll was used by Thomas Thomas, later University Printer.
[2] I am indebted to Mr N. Ker for knowledge of this roll. See p. 49, n. 3.

HM. *h* (15)–
MW. *d* (9)

HM. *h* FOLIAGE, RENAISSANCE, OR NON-
DESCRIPT ORNAMENT (*cont.*)

 With HM. *h* (1) 1534
 With ornament B (9) 1530–2
 With ornament C (5) 1530

(16)

 With ornament C (1) 1519
 With ornament B (2) 1524

(17) London
 Alone 1562
 With HM. *g* (5) 1524

(18) ? English[1]
 Alone 1586

(19) G.C. London
 Alone 1567, 1579

(20) London
 Alone 1541–60
 With HE. *b* (5) ?1585–1588
 With FP. *g* (5) 1553
 With RP. *c* (2) ?1571
 With SW. *b* (5) ?1577
 With HM. *b* (5) ?1585
 With HE. *k* (1) ?1585
 With ornament J (4) 1552

(21) London
 With FP. *f* (5) ?1591–?1596
 With DI. *d* (2) ?1591–?1596

(22) ? English[1]
 Alone 1589–91

(23) London
 Alone 1572

(24)

 Alone 1556
 With HM. *a* (17) 1551–64

(25)

 Alone 1535–56

(26)

 Alone 1535, 1567–71

(27) ? English
 Alone 1538

(28) London
 Alone 1526–41
 With HM. *h* (29) 1523–36
 With FL. *a* (1) 1529–45

(29) London
 Alone 1533–44
 With HM. *h* (28) 1523–36
 With FL. *a* (1) 1536

(30)

 Alone 1545–?1573

(31)

 With HM. *e* (5) 1546

MW. METALWORK. (PL. LI)

MW. *a* FLEUR-DE-LIS IN CENTRE

(1) G.K. Oxford
 Alone 1535–1621
 With FP. *g* (9) 1513–54, 1569–1605
 With IN. (3) 1540
 With HM. *a* (1) [G.F.] 1546

With DI. *a* (5) 1546
With HM. *d* (1) 1559
With HM. *h* (4) 1473
With IN. (11) 1604

(2) N.S. Oxford
 Alone 1573–1622

(3) P.L.
 Alone 1546–55

(4) London
 With HE. *k* (1) 1550–62

MW. *b* ROUNDEL IN CENTRE

(1) Oxford
 Alone 1550–60
 With RP. *d* (1) 1531–54
 With RC. *c* (1) 1562–7
 With FC. *c* (1) 1533
 With FP. *g* (9) 1544
 With HM. *d* (1) 1527, 1561

MW. *c* BAR IN CENTRE

(1) London
 Alone 1555–71
 With HE. *b* (5) 1559–67

MW. *d* FLOWER IN CENTRE

(1) R.P. London
 Alone 1564–6
 With FP. *g* (5) 1512, 1551–70
 With HE. *b* (6) 1570

(2) London
 Alone 1536–56
 With MW. *d* (5) [G.M.] 1534–52
 With HE. *g* (3) [H.R.] 1576
 With FP. *f* (5) 1576
 With FP. *g* (11) 1576

(3) London
 Alone 1522, 1540–63, ?1668

(4)

 Alone 1546–69

(5) G.M. London
 Alone 1543–9
 With MW. *d* (2) 1534–52

(6) London
 Alone 1551
 With HM. *a* (6) 1529–56

(7) London
 Alone 1540, 1550
 With HM. *h* (7) [F.P.] 1549

(8) London
 Alone 1554
 With HE. *g* (2) [H.R.] 1550–5
 With HE. *k* (1) 1550

(9) Oxford
 Alone 1528–53
 With HM. *h* (4) 1532–50
 With HM. *d* (1) 1537
 With FC. *c* (1) 1537

---

[1] This has a rather Spanish appearance, and may not be English.

MW. *d.* FLOWER IN CENTRE (*cont.*)

(10) Oxford
Alone ? 1560–1613
With MW. *d* (11) 1535–57
With FL. *c* (1) 1572–?1605
With HM. *h* (2) 1586–?1605
With IN. (11) 1596

(11) Oxford
Alone 1613–14
With MW. *d* (10) 1535–57
With IN. (2) 1534–46
With DI. *a* (5) 1546
With FP. *a* (10) 1551
With FP. *g* (6) [R.W.] 1610
With FP. *g* (7) [I.A.] 1613

(12) London
Alone 1549–52
With HM. *h* (10) [R.B.] 1550–70
With HE. *b* (5) 1555–70
With HM. *a* (6) 1527, 1562

(13) London
Alone ? 1567
With GE. (6) 1572

(14) London
Alone 1557
With HM. *a* (13) 1531

(15) ? English
Alone 1558

RC. RENAISSANCE: CONTINUOUS.
(PL. LII)

RC. *a* CAPSTAN

(1) London
Alone 1537
With HM. *b* (3) [F.D.] 1596
With FC. *a* (3) 1596
With HE. *d* (1) 1596

(2) Z.C. ? Cambridge
Alone 1522
With FL. *d* (1) 1519, 1529
With AN. *h* (1*a*) 1519

RC. *b* VASES, ETC.

(1) Oxford
Alone 1546–65
With FL. *a* (7) 1562
With FP. *g* (9) 1564

(2) London
With HM. *f* (1) 1515–25
With HM. *c* (1) n.d.
With CH. *c* (10) n.d.

(3) R.W. London
Alone 1514, 1528–37
With FP. *a* (4) 1520–41
With CH. *a* (1) 1523–39
With FC. *e* (1) 1534
With FL. *a* (7) 1539

(4) N.E. London
Alone 1545, ? 1579
With FP. *a* (3) 1537–45
With FL. *a* (11) 1571–3
With HE. *a* (1) ? 1586

(5) ? London
With HM. *h* (15) 1530–1
With RC. *b* (6) 1531
With ornament C (5) 1530

(6) London
Alone 1508–34
With FP. *g* (2) 1530
With TP. (3) 1530
With AN. *g* (1) 1530
With HM. *h* (15) 1531
With RC. *b* (5) 1531
With HM. *a* (7) 1536–8
With HE. *c* (1) 1544
With HM. *c* (6) n.d.
With rose panel: n.d.
With animals in foliage panel: n.d.

(7)
With RC. *b* (8) n.d.

(8)
With RC. *b* (7) n.d.

RC. *c* DATED

(1) Oxford
Alone 1531–73
With DI. *a* (5) 1520–45
With HM. *a* (1) [G.F.] 1528–44
With RC. *d* (1) 1532–51
With HM. *a* (5) 1537
With MW. *b* (1) 1562–7
With FP. *g* (9) 1564
With ornament J (3) ? 1537

RC. *d* HEADS

(1) London and Shrewsbury
Alone 1486, 1525
With SV. *a* (2) 1548, 1564

(2) London
Alone 1529

RC. *e* NONDESCRIPT

(1) Oxford and probably London
Alone 1512
With FL. *a* (4) 1519–30
With MU. (2) 1520–30
With FF. (1) ? 1524–9
With MU. (1) 1532
With ornament B (5) 1509–31
With ornament A (4) 1519
With ornament C (1) 1503, 1533–6
With stamp 1063: 1519
With Royal Arms and rose panels (Hobson, *Panels*, pp. 32–3, III*a* and *b*) 1506

RP. RENAISSANCE: PANELLED.
(PL. LIII)

RP. *a* CAPSTAN

(1) London
Alone 1525–59
With HE. *a* (1) 1549–54
With FL. *a* (5) 1533
With stamp 1034: 1523
With stamp 1055: 1523

MW. *d* (10)–
RP. *a* (1)

RP. *a* (2)–
TC. *a* (8)

RP. *a* CAPSTAN (*cont.*)

(2) London
Alone 1512–51
With SW. *b* (3) 1532–43
With SW. *b* (4) 1534–6, 1573
With HE. *g* (4) 1550
With IN. (10) 1550

(3) London
Alone n.d.
With CH. *a* (5) n.d.
With FP. *a* (18) n.d.

(4) London
Alone 1538–50
With HM. *a* (10) 1532–49

(5) London
Alone 1538–52
With FP. *a* (6) [F.I.] 1538–51
With HM. *a* (7) 1523, 1531
With FL. *a* (5) 1538
With CH. *c* (6) 1550
With DI. *e* (2) 1555

RP. *b* BIRDS AFFRONTÉS AND ADDORSED

(1) London
Alone 1539–44
With TP. (1) 1524–48
With HM. *a* (4) 1504
With HM. *h* (11) 1549
With FC. *a* (2) 1561

(2)
Alone 1535

RP. *c* BIRDS AFFRONTÉS

(1) Oxford
Alone 1557–1618
With FL. *a* (7) 1559–64
With HM. *h* (3) 1562–4
With FL. *c* (1) 1598–1630
With IN. (7) 1586

(2) ?London
Alone 1535
With SW. *b* (6) ?1556
With HM. *h* (20) ?1571

RP. *d* HALF-WHEELS

(1) Oxford
Alone 1536–1616
With MW. *b* (1) 1531–54
With RC. *c* (1) 1532–51
With DI. *a* (5) 1532–51
With FP. *g* (9) 1544–55
With FC. *c* (1) 1533, 1550
With FP. *g* (7) [I.A.] 1598–1612
With HM. *h* (4) ?1605

(2)
With CH. *c* (4) 1550

RP. *e* COLUMNS, WHEELS

(1) London
Alone 1550–6

RP. *f* NONDESCRIPT

(1) London
With AN. *e* (2) [W.G.-I.G.] 1530–3
With stamp 37: 1531
With St Bernard panel (Hobson, *Panels*,
p. 55, no. 2) 1529
With animals in foliage panel (Weale,
R. 99) 1530
With acorn panel (Weale, R. 99) 1530

(2) London
Alone 1528
With FL. *a* (6) 1493
With HE. *a* (1) 1540
With Royal Arms and rose panels,
signed H.I. (Hobson, *Panels*, p. 39,
II, IIIb) 1531

(3) London
Alone 1525

(4) London
Alone 1542
With AN. *b* (1) [I.R.] 1532–40
With Reynes' signed panels (Weale,
R. 125, 126) n.d.

CR. CRESTING. (PL. LIV)
CR.

(1) London
With AN. *b* (3) 1514

(2)
Alone 1532
With HM. *d* (4) 1528
With DI. *ḣ* (3) 1530

(3) T.M. London
With DI. *a* (13) 1609

(4) E.B. London
Alone 1493
With HE. *l* (1) 1610

TC. TUDOR CRESTED. (PLS. LIV, LXI)
TC. *a* UNSIGNED

(1) London
Alone 1634

(2) London
Alone 1595
With HM. *g* (1) 1506, 1577–84, 1599–
1627, ?1647

(3) London
With GE. (1) 1571, 1584
With HE. *l* (1) 1613–23

(4) London
Alone 1543, 1611

(5) London
With FP. *b* (3) ?1562, 1581, ?1615

(6) London
With HM. *h* (14) [R.D.] 1499
With HE. *l* (3) [E.R.] 1592
With SV. *a* (3) 1623

(7) London
With HE. *b* (5) n.d.
With GE. (1) 1609

(8) London
Alone 1495

TC. *b* Signed

  (1) T.E. London
      With HE. *l* (2) 1613
  (2) I.P. (initials under rose) London[1]
      Alone n.d.
  (3) H.C. London
      With FP. *g* (13) ?1634
  (4) I.P. (initials within a heart) London
      Alone 1632
  (5) P.S.H. London
      Alone 1625

SV. SIDE-VIEW. (PL. LV)

  SV. *a* Continuous

  (1) Cambridge and London
      Alone 1511, 1520, 1544
      With HE. *b* (4) [I.S.] 1522
      With ornament A (1) 1511–15
      With ornament I (2) 1504
      With ornament K (7) 1512
      With stamp 1054: 1514–16
  (2) London and Shrewsbury
      Alone 1493, 1538, 1550, 1612
      With RC. *d* (1) 1548, 1564
      With stamp 1045: n.d.
      With stamp 1046: n.d.
      With Royal Arms and rose panels signed
        H.N. (Hobson, *Panels*, p. 39 *d*) n.d.
  (3) London
      Alone ?1496
      With FP. *g* (12) 1556
      With FP. *g* (3) 1568
      With TC. *a* (6) 1623
      With ornament D (1) 1506
  (4) London
      Alone ?1624
  (5)
      Alone ?1509
      With ornament I (3) 1504–19
      With stamp 320: 1519–26
      With stamp 321: 1519–26
      With Coronation of the Virgin panel
        signed R. Macé (Weale, R. 97) 1520
      With Annunciation panel (Weale, R.
        97) 1520
      With Tudor rose panel (Weale, R. 96)
        1519–26
      With double eagle panel (Weale R. 96)
        1519–26
  (6) K.L.-L.K. London
      Alone 1516–26
      With AN. *b* (3) 1516–28
      With DI. *a* (8) 1516–28
      With TP. (2) n.d.
      With FF. (2) 1520
      With ornament B (1) 1524
      With rose and scroll panel, signed I.N.
        (Hobson, *Panels*, p. 39, *e*) 1526

      With Royal Arms panel (Hobson,
        Panels, p. 39, *e*) 1526
  (7) R.H.M.I.
      Alone 1497–1523
      With IN. (4) 1520
      With ornament A (4) n.d.
      With ornament H 9 (*c*) 1488
      With stamp 1058: 1519
      With stamp 1064: 1519
  (8) London
      With HE. *b* (5) 1580–4

  SV. *b* Panelled

  (1) London
      Alone n.d.
  (2)
      With HE. *a* (1) ?1585
      With HE. *b* (5) ?1585

IN. INTAGLIO. (PLS. LV, LXI)

  IN.

  (1) London
      Alone 1572
      With HM. *h* (10) 1573
  (2) Oxford and ?London
      With MW. *d* (11) 1534–46
      With rose panel[2] ?1522
  (3) Oxford
      Alone 1581–1617
      With HM. *h* (4) 1532
      With MW. *a* (1) [G.K.] 1540
  (4) ?Oxford, and probably London
      With CH. *c* (1) 1527–36
      With HM. *h* (1) 1534–41
      With SV. *a* (7) 1520
  (5) Norwich
      With FC. *h* (12) 1548
      With FP. *g* (13) ?1561
  (6) London
      Alone 1536
      With HM. *h* (10) [R.B.] 1578
  (7) Oxford
      Alone 1504, 1562–98, 1609–18
      With RP. *c* (1) 1586
  (8) London
      With HM. *h* (5) [I.R.] 1568
  (9) ? Cambridge[3]
      With HM. *h* (14) 1544
      With HM. *g* (2) 1577
  (10)[4]
      With HE. *g* (4) 1550
      With RP. *a* (2) 1550
      With GE. (3) 1575
  (11) Oxford
      With MW. *d* (10) 1596

TC. *b* (1)–IN. (11)

---

[1] The roll as reproduced is not complete.

[2] This is one of the numerous variants of Pynson's rose panel. This example is on Brit. Mus. C. 69 ff. 8. The only other example I know is no. 1323 in Mr A. Ehrman's collection, where it is used with another variant.

[3] Mr J. C. T. Oates tells me that this roll was used by Thomas Thomas, later University Printer.

[4] I am indebted to Mr N. Ker for knowledge of this roll.

IN. *(cont.)*
>>> With HM. *h* (2) 1603
>>> With MW. *a* (1) 1604

SW. STRAPWORK. (PL. LVI)
SW. *a* FRENCH ROYAL EMBLEMS
(1)
>>> With AN. *d* (1) 1532
>>> With stamp 1033: 1518
>>> With stamp 1038: 1518

SW. *b* FOLIAGE OR NONDESCRIPT ORNAMENT
(1) London
>>> With CH. *a* (1) 1531
(2) London
>>> With HM. *a* (4) 1523–31, 1552–9, ?1600
>>> With HM. *a* (7) 1523–31, 1559
>>> With HM. *e* (2) ?1600
>>> With FC. *g* (1) ?1600
>>> With HE. *a* (1) ?1600
(3) London
>>> Alone 1511–40
>>> With FP. *a* (7) ?1516–1539
>>> With TP. (1) ?1522–1534
>>> With RP. *a* (2) 1532–43
>>> With AN. *e* (2) [W.G.-I.G.] ?1522
>>> With FL. *d* (1) 1529
>>> With SW. *b* (4) 1530–3
>>> With FP. *a* (20) 1535
>>> With ornament E (1) 1522
(4) London
>>> With SW. *b* (3) 1530–3
>>> With RP. *a* (2) 1534–6, 1573
(5) London
>>> Alone 1559–83
>>> With HE. *g* (2) [H.R.] 1553–74
>>> With CH. *c* (9) 1567–70
>>> With HM. *a* (6) 1568–73
>>> With HM. *b* (3) [F.D.] 1570–4
>>> With HM. *h* (20) ?1577
(6)
>>> Alone ?1508–1533, 1561
>>> With TP. (3) 1534
>>> With FP. *g* (2) 1534
>>> With RP. *c* (2) ?1556
(7) London
>>> Alone 1556

MU. MUSICAL INSTRUMENTS.
(PL. LVI)
MU.
(1) Oxford, and probably London
>>> Alone 1520–34
>>> With RC. *e* (1) 1532
>>> With ornament C (1) 1518–35
>>> With ornament H 9 (*a*) 1528
(2) Oxford and probably London
>>> With RC. *e* (1) 1520–30
>>> With FL. *a* (4) 1521–5
>>> With FF. (1) 1529
>>> With ornament B (5) 1529
>>> With stamp 1039: 1523–5
>>> With St James panel 1524

TP. TWISTED PINEAPPLES. (PL. LVI)
(1) London
>>> Alone 1533
>>> With SW. *b* (3) ?1522–1534
>>> With RP. *b* (1) 1524–44
>>> With AN. *e* (2) [W.G.-I.G.] ?1522
>>> With FC. *h* (10) [R.C.] n.d.
(2) London
>>> Alone 1519
>>> With HM. *a* (4) ?1516
>>> With FP. *a* (9) ?1516
>>> With SV. *a* (6) [K.L.-L.K.] n.d.
(3)
>>> With AN. *g* (1) 1530
>>> With RC. *b* (6) 1530
>>> With SW. *b* (6) 1534
>>> With FP. *g* (2) 1534

FF. FULL FIGURES. (PL. LVI)
(1) De Villiers, London
>>> With RC. *e* (1) ?1524–1529
>>> With MU. (2) 1529
>>> With ornament A (1) 1493, 1511
>>> With ornament B (5) 1529
(2)
>>> With SV. *a* (6) [K.L.-L.K.] 1520
(3)
>>> Alone 1537–52

CLASSIFIED LIST OF ORNAMENTS

A. PINEAPPLE. (PL. LVII)
(1) London
>>> With AN. *b* (1) [I.R.] 1479, 1495, 1503–40
>>> With AN. *b* (2) 1482, 1488, 1491–1520
>>> With FF. (1) [De Villiers] 1493, 1511
>>> With SV. *a* (1) 1511–15
(2) Cambridge and London
>>> With AN. *h* (1) [L.V.L.] 1502–9
>>> With HE. *d* (1) 1505–27
>>> With DI. *b* (1) 1501
(3)
>>> With AN. *h* (2) [H.C.] 1508, 1515

(4) London and probably Oxford
>>> With AN. *b* (3) 1497
>>> With RC. *e* (1) 1519
>>> With FL. *a* (4) 1521
>>> With SV. *a* (7) n.d.
(5)
>>> With DI. *a* (6) 1501, 1512
(6) London
>>> With FP. *g* (2) 1517
(7) Oxford
>>> With stamps used by Fruit and Flower
>>> Binder 1486–98

A. PINEAPPLE (*cont.*)

(8)
    Alone 1510
(9) London
    With FL. *a* (4) 1522
    With stamp 1063: 1522
(10)
    With stamps used by Binder F. 1488–95

B. PINEAPPLE WITH BLOB BELOW.
    (PL. LVII)

(1) London
    With AN. *b* (3) 1511–23
    With DI. *c* (1) 1488, 1504
    With SV. *a* (6) 1524
(2)
    With HM. *d* (2) 1518–28
    With HM. *h* (16) 1524
    With AN. *i* (1) 1529
(3) London
    With AN. *g* (1) 1496, 1500–5, 1514–34
    With AN. *e* (1) [W.G.–I.G.] ? 1514
(4) ? London
    With FL. *b* (3) 1522
    With HM. *a* (22) 1532
    With CH. *c* (8) 1532
(5) Oxford and London
    With FL. *a* (4) 1492, ? 1494, 1508–27
    With RC. *e* (1) 1509–31
    With FF. (1) 1492, 1529
    With AN. *h* (1) [L.V.L.] 1502
    With AN. *m* (1) 1505
    With MU. (2) 1529
    With stamp 399: n.d.
(6) ? Oxford, ? London
    With AN. 1 (1) 1502
    With stamp 182: 1506–9
    With stamp 1028: 1496, 1506
(7)
    With HM. *d* (2) 1516
    With FC. *c* (2) 1516
(8)
    With stamps used by the Lily Binder 1486, 1504
(9)
    With HM. *h* (15) 1530–2
(10) ? London
    With FL. *b* (8) 1508

C. CRUCIFORM. (PL. LVIII)

(1) ? London
    With MU. (1) 1518–35
    With HM. *h* (16) 1519
    With FL. *a* (4) 1526
    With HM. *a* (5) 1532
    With RC. *e* (1) 1503, 1533–6
    With HM. *h* (1) 1536
(2) London
    With AN. *k* (1) n.d.
(3)
    With DI. *a* (6) 1515

(4) London
    With HE. *h* (1) 1511–35
    With HE. *g* (4) 1526–34
(5) London
    With HE. *a* (1) 1521–30
    With FL. *b* (8) 1516
    With HM. *h* (15) 1530
    With RC. *b* (5) 1530
    With AN. *b* (1) [I.R.] 1531

D. ACORNS. (PL. LVIII)

(1) London
    With DI. *a* (6) 1502–16
    With SV. *a* (3) 1506
(2) London
    With AN. *b* (3) 1508–14
    With FR. (3) 1504

E. STRIPED. (PL. LVIII)

(1) London
    With AN. *e* (1) [W.G.–I.G.] 1506–25
    With AN. *e* (2) [W.G.–I.G.] 1511–26
    With SW. *b* (3) 1522

F. FLEUR-DE-LIS SHAPE. (PL. LVIII)

(1)
    With AN. *b* (3) 1505
    With AN. *m* (1) n.d.
(2) London
    With AN. *d* (1) 1528
    With stamp 37: 1528
    With stamp 235: 1528

G. PINEAPPLE WITH PLAIN
    CENTRE. (PL. LVIII)

(1) Probably Oxford
    With stamps used by the Greyhound
        Binder, n.d.
(2) Probably Oxford
    With stamps used by the Greyhound
        Binder, n.d.

H. LATTICE. (The number given to each
    group indicates the number of perforations.
    PL. LVIII)

9. (*a*) Oxford
    Alone 1510
    With FL. *a* (3) 1500–17
    With MU. (1) 1528
    With stamps used by the Fruit and Flower
        Binder 1492
    With dragon stamp (Hobson, *Cambridge*,
        Pl. X, no. 13) 1494
9. (*b*) Oxford
    With FP. *f* (4) [D.V.] ? 1493
    With stamps 420, 421: 1494
9. (*c*) Oxford
    With FL. *a* (3) 1513–15
    With SV. *a* (7) 1488
    With AN. 1 (1) n.d.
    With stamp 322: n.d.

H. (**13**)–K. (7)

H. LATTICE (*cont.*)

13.　(*a*) Cambridge
　　　Alone 1500–20
　　　With HE. *b* (2) [G.G.] 1505–12
　　　With DI. *a* (1) 1506–13
　　　With stamps used by the Lattice Binder,
　　　　1485–1510

16.　(*a*) Cambridge
　　　With stamps used by the Unicorn, Lattice,
　　　　Heavy and Monster Binders 1495–
　　　　1505

16.　(*b*)
　　　With FR. (3) 1510
　　　With AN. *m* (1) 1521
　　　With stamp 1056: 1521

16.　(*c*)
　　　With FR. (3) ?1512
　　　With stamp 1047: 1512

18.　(*a*)
　　　With foliage border stamp, 1497
　　　With stamp 333: n.d.

20.　(*a*)
　　　With AN. *h* (2) [H.C.] 1496, 1508
　　　With AN. *h* (1) [L.V.L.] 1497
　　　With FR. (1) 1502
　　　With DI. *b* (1) 1524
　　　With stamps used by the Huntsman Binder
　　　　1490–8, 1524
　　　With stamp 1029: 1502
　　　With stamp 30: 1502

20.　(*b*)
　　　With stamp 1050: 1486

HS. HALF-STAMPS. (PL. LIX)

(1)
　　　Alone 1500
　　　With stamps used by the Half-stamp
　　　　Binder 1489–1504

(2)
　　　Alone 1514, 1524
　　　With AN. *h* (2) [H.C.] 1513
　　　With FL. *a* (13) 1519, 1528
　　　With FL. *b* (6) 1526

(3)
　　　With stamps used by the Half-stamp Binder
　　　　1493

(4)
　　　With DI. *a* (6) 1520

I. FRINGED FOLIAGE. (PL. LIX)

(1)
　　　With AN. *h* (2) [H.C.] 1503
　　　With stamps used by the Foliaged Staff
　　　　Binder 1497, 1500

(2) Cambridge and London
　　　With AN. *h* (2) [H.C.] 1496–?1502
　　　With AN. *g* (1) 1504–12
　　　With SV. *a* (1) 1504
　　　With HE. *b* (2) [G.G.] 1511

(3) Cambridge and ?London
　　　With SV. *a* (5) 1504–19
　　　With stamps used by the Unicorn and
　　　　Monster Binders 1481–94
　　　With stamp 399: 1501–7

(4)
　　　With stamps used by the Foliaged Staff
　　　　Binder 1492
　　　With stamp 267: 1500
　　　With eagle and lily stamps 1500

(5)
　　　With stamps used by the Lily Binder 1498

J. HALF-OVALS. (PL. LIX)

(1)
　　　Alone n.d.

(2)
　　　With HM. *a* (8) 1521

(3) Oxford
　　　With HM. *a* (5) ?1537
　　　With RC. *c* (1) ?1537
　　　With FL. *f* (1) n.d.

(4) London
　　　With HM. *h* (20) 1552

K. NONDESCRIPT. (PL. LIX)

(1) London
　　　With FC. *b* (2) 1513–23

(2) London
　　　With AN. *b* (1) 1516

(3) Cambridge
　　　With HE. *b* (3) 1502–15
　　　With AN. *f* (2) 1507–15
　　　With DI. *a* (4) 1519
　　　With stamps used by Spierinck 1507–15

(4) Cambridge
　　　Alone 1491–1516
　　　With HE. *b* (3) ?1493, 1508–13
　　　With DI. *a* (4) 1505–12
　　　With stamps used by Spierinck 1492–1521

(5)
　　　With stamps 1022, 1036: 1501

(6)
　　　Alone 1496

(7) London
　　　With AN. *b* (2) 1482, 1491, 1498–1506
　　　With AN. *b* (1) [I.R.] 1520–32
　　　With SV. *a* (1) 1512

INDEX TO PLATE LX

INDICATING UNDER WHAT ROLLS OR ORNAMENTS IN THE
FOREGOING LISTS EACH STAMP WILL BE FOUND

1021 AN. *g* (1)
1022 K. (5)
1023 AN. *b* (3), FL. *a* (12)
1024 HE. *j* (2)
1025 AN. *j* (1), DI. *a* (7),
     FP. *b* (1), FL. *a* (4)
1026 FR. (3)
1027 AN. *h* (1*a*)
1028 B. (6)
1029 FR. (1)
1030 FL. *a* (4)
1031 DI. *b* (1)
1032 FP. *f* (4)
1033 SW. *a* (1)
1034 RP. *a* (1)
1035 AN. *h* (1)
1036 K. (5)
1037 AN. *d* (1), DI. *b* (1)
1038 SW. *a* (1)
1039 FL. *a* (4)
1040 FL. *a* (4)
1041 AN. *i* (1)
1042 DI. *c* (1), FC. *b* (2)
1043 DI. *c* (1)
1044 DI. *c* (1), FC. *b* (2)

1045 SV. *a* (2)
1046 SV. *a* (2)
1047 FR. (3)
1048 AN. *d* (1), AN. *e* (2),
     FP. *f* (4), RP. *f* (1)
1049 FP. *f* (4)
1050 H. 20 (*b*)
1051 AN. *h* (1)
1052 AN. *j* (1)
1053 AN. *c* (1), AN. *k* (1),
     FC. *b* (2)
1054 AN. *h* (1*a*), SV. *a* (1)
1055 RP. *a* (1)
1056 AN. *m* (1)
1057 FL. *a* (15)
1058 SV. *a* (7)
1059 AN. *g* (1)
1060 FL. *a* (4)
1061 AN. *m* (1)
1062 AN. *d* (1), DI. *b* (1)
1063 FL. *a* (4), RC. *e* (1)
1064 SV. *a* (7)
1065 AN. *b* (3)
1066–73 FC. *h* (6)
1074 AN. *h* (1)

GLOSSARY OF TERMS

USED IN CONNEXION WITH BLIND-STAMPED BINDINGS

ACANTHUS: an ornament, frequently used on heads-in-medallions rolls, which appears to represent two acanthus leaves pointing different ways (see fig. 1).

ANIMALS IN FOLIAGE PANEL: the characteristic Netherlandish design in which a panel is divided vertically into two, each section containing curving foliage with an animal within each curve.

BACK: the part of the binding, sometimes called the spine, which joins the two covers.

BLANK BOOK: a book bound up with blank leaves only, for entering accounts, records, etc.

BLIND-STAMPED BINDING: any leather binding decorated by stamping, whether by hand or in a press, without the use of gold.

BOARDS: the stiff material of the covers, whether wood, cardboard or sheets of paper stuck together. On rare occasions they were of leather.

BORDER: decoration that actually borders the edges of the cover. Cf. 'Frame'.

BRIQUET: a conventional representation of a steel for striking a light (see fig. 2).

BUDS: an ornament filling a small panel on some rolls, which is a conventionalised form of a spray bearing buds (see fig. 3).

CAPSTAN: an ornament, common on English and French heads-in-medallions rolls, roughly resembling a capstan (see fig. 4).

CARTOUCHE: a small rectangular ornament, used on rolls, formed by one or more lines, generally with a plain centre (see fig. 5).

CATCH: the metal catch and its attachment, over which the clasp fits. In some cases it is a pin, not a bar.

CENTREPIECE, CORNERPIECE: a piece of metal, usually embossed and engraved, lozenge-shaped when used in the centre, square (or roughly so) when used at the corners, fastened on to the covers.

CENTRE-STAMP, CORNER-STAMP: a piece of decoration for respectively the centre or corner of a cover, produced by a single tool used in a press.

CHEMISE: an extension of the leather covering of the boards, either to protect the edges, or, in some cases, to enable the book to be attached to the owner's girdle.

CLASP: the metal fitting, or the thong to which it is attached, which slips over the catch.

63

CLIP:[1] the metal eye, of whatever form, fixed to one of the covers, to which the book's chain was attached.

CONVENTIONAL FOLIAGE: ornament, often quite unrealistic, but evidently suggested by foliage (see fig. 6).

CRESTING: used of a roll consisting of two undulating and intersecting lines below, and above a series of tufts (see fig. 7).

CROCKETED CRESTING: a frame formed by roughly rectangular stamps ornamented with crockets, or by roughly triangular stamps which, set together and pointing outwards, give a kind of cresting effect (see Pl. V). Rarely a roll is used for the same purpose (e.g. Roll CR. (4)) (see figs. 8, 9).

DIAPER: a small pattern indefinitely repeated, often in the form of lozenges and triangles containing quatrefoils or conventional flowers.

DOUBLURE: a decorated leather lining in place of an ordinary pastedown. This is rare in blind-stamped bindings, but common in gold-tooled bindings, when the doublure may be of leather (sometimes vellum) or silk.

ENDPAPERS: all the paper added to a book by the binder, including both pastedowns and fly-leaves. Often part of the endpapers is not actually of paper, but of vellum, and rarely of other leathers.

FEATHER ORNAMENT: engraved ornament on clasps or catches imitated from feathers.

FERN-TIP: ornament resembling a row of tips of ferns, less common in blind than in gold work (see fig. 10).

FILLET: a line, whether single, double or triple.

FLEUR-DE-LISÉ LOZENGE: a lozenge stamp consisting of a flower with fleurs-de-lis filling the corners, or a variant of this design (see fig. 11).

FLEURON: any nondescript ornament, partly of floral or foliage character, generally of roughly lozenge shape, that cannot be more precisely designated, used ordinarily to fill the lozenge compartments of a panel that is divided up by diagonal fillets (see fig. 12).

FLOWER-HEADED RIVET: a rivet with an ornamental head of roughly daisy design.

FLY-LEAF: any free leaf added by the binder at either end of a book. See under 'Endpapers'.

FOLIAGED STAFF: a branch entwined with foliage, sometimes accompanied by berries; such a stamp is generally used repeated to form a frame (see fig. 13).

FOUNT: the whole collection of tools (see below) used by a single binder.

FRAME: decoration running parallel to the four sides of the cover, but with a space between it and the edges. Cf. 'Border'.

FREE: used of a stamp, of whatever form, which has no bounding line round it; such stamps may be in intaglio or relief.

FRINGED FOLIAGE ORNAMENT: a roughly lozenge-shaped ornament of conventional foliage, the characteristic feature of which is a shallow fringe round its edge (see fig. 14).

GAUFFRED: applied to the decoration of the edges by means of tooling, as opposed to painting.

GRID: an ornament frequently used on heads-in-medallions rolls, consisting of two horizontal lines with a few short vertical bars between them, the sides having a foliage character (see fig. 15).

GROWING FLOWER: a common ornament on rolls, consisting usually of a flattened elliptical base, from which springs a stem bearing leaves, and, at the top, two flowers whose tops curl outwards (see fig. 16).

HALF-BANDS: bands forming a small ridge across the back of a book that has also full bands, being usually produced by the use of cord, as opposed to leather, or of one strip of cord or leather as opposed to two; sometimes they merely mark the position of the kettle-stitch.

HALF-STAMPS: stamps whose design is the same as, or similar to, one half of a fleuron, pineapple or lattice stamp, etc.; they are used ordinarily to fill the triangular compartments of a panel that is divided up by diagonal fillets, the stamp being divided vertically for the side compartments, and horizontally for those at top and bottom; sometimes they are doubled to fill a lozenge compartment (see HS. on Pl. LIX).

---

[1] This term has the authority of the Bodleian Day Book, 1613–14 (Gibson, pp. 51–5).

HATCHING: a row of diagonal lines slanting the same way, found sometimes on the backs or the head and tail edges of books; cross-hatching: the same, except that the lines cross one another in opposite directions.

HEADBAND: specifically, the silk band at the head; in plural, loosely for head- and tailbands.

HERALDIC CRESTING: cresting (see above) the projections of which terminate in heraldic (usually Tudor) emblems.

INTAGLIO: used of a roll or stamp in which the design, *as impressed on the leather*, appears in intaglio.

INTERSECTING FRAME: a frame the sides of which (or some of them) are extended, where they meet each other, to the edges of the cover.

LATTICE STAMP: an ornament the distinguishing feature of which is a central diamond formed of lattice or criss-cross work (see fig. 17).

LOZENGE: a four-sided stamp or compartment standing on one of its points, whether square or not.

MERRYTHOUGHT: a stamp of the form of a merrythought, usually decorated with cusps or foliage ornament (see fig. 18).

METALWORK: an ornament filling a small panel on many rolls (especially those used at Oxford), clearly imitated from wrought and curved ironwork (see fig. 19).

MITRED: used of the connexion at the angles of an outer frame to an inner frame or panel by the diagonal use of fillets or a roll.

PANEL: (apart from the non-technical use of the term) a tool (or its impression), generally rectangular, used in a press to produce a design or picture usually much larger than a stamp (see below).

PASTEDOWNS: those endpapers with which the covers are lined, whether of paper or vellum, plain, printed or manuscript.

PINEAPPLE: an ornament bearing some likeness to a conventional pineapple, used in the same position as a fleuron (see fig. 20).

REINFORCING-PIECE: the paper or parchment on which, for greater strength, the backs of some or all of the sections are often sewn, part of which shows under the pastedowns.

RELIEF: used of a roll or stamp in which the design, *as impressed on the leather*, appears in relief.

RENAISSANCE ORNAMENT: conventional ornament apparently suggested by columns, urns, vases, beasts, birds, garlands and foliage such as appear in the ornament of Renaissance architecture (see fig. 21).

RIGHT, LEFT: as seen by the spectator.

ROLL: a tool (or its impression) having a continuous design engraved on a wheel.

SCROLL: a scroll-shaped stamp used for bearing an inscription, or an ornament of rather similar shape, used generally between flowers on a roll (see FL. *d* on Pl. XLIV).

SERRATED SQUARE: a stamp, more or less square, with concave serrated sides, and usually a cruciform centre; often used in a group giving the effect of a number of circles with serrated inner edges (see fig. 22).

SHOES: metal attached to the edges only of the covers at the corners, and sometimes next the back at the tail, to protect the leather.

STAMP: a tool (or its impression) used by hand pressure to produce a single ornament of any kind; also used loosely in such phrases as panel stamp, armorial stamp, centre-stamp, corner-stamp, though in these cases, the impression would be made in a press, not by hand.

STRAPWORK: double lines, suggesting a strap, generally used interlacing (see fig. 23).

TAILBAND: the silk band at the tail.

TOOL: (as applied to finishing) the actual die, usually of brass, used by the binder, whether in the form of a stamp, a roll, or a panel.

TURN-IN: the portion of leather that shows along the edges on the inside of the covers.

TWISTED PINEAPPLES: the form of ornament used on a number of English and French rolls, consisting of twisted stems with conventional pineapples at intervals (see TP. on Pl. LVI).

UPPER COVER, LOWER COVER: respectively the cover next to the title or first page, and the cover at the other end.

VOLUTE: an ornament, frequently used on heads-in-medallions rolls, consisting of a large curl in roughly volute form at one end, and at the other a small curl turning the opposite way; always used in pairs (see fig. 24).

WASTE, BOOKSELLER'S: any fragment of books, e.g. excess quires of a recently printed book, or remains of old books, manuscript or printed, treated as worthless and used up by a bookseller or binder. Cf. Goldschmidt, p. 119.

WASTE, PRINTER'S: printer's trial sheets printed on one side only, or other printed pages thrown away by the printer, and used up by a bookseller or binder. Cf. Goldschmidt, p. 119.

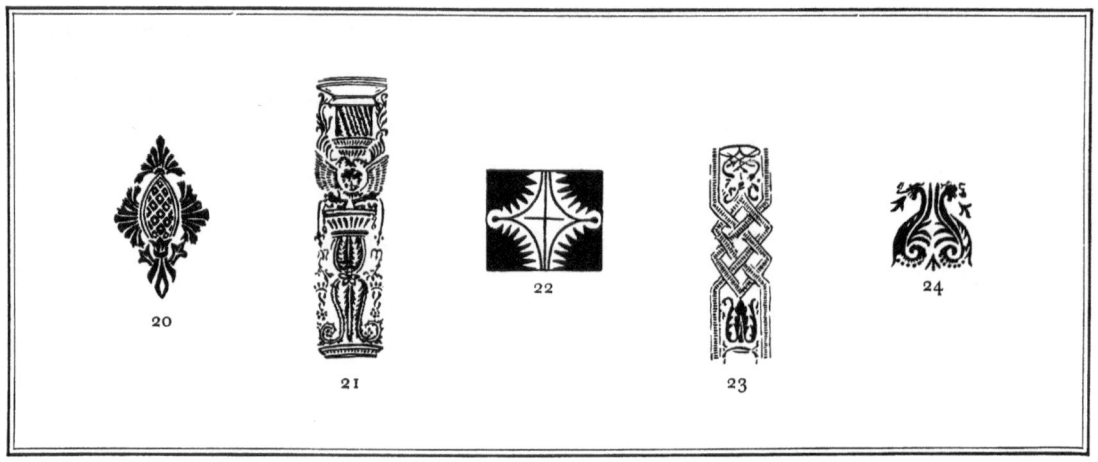

INDEX

INDEX

INDEX

INDEX

INDEX

PLATE I

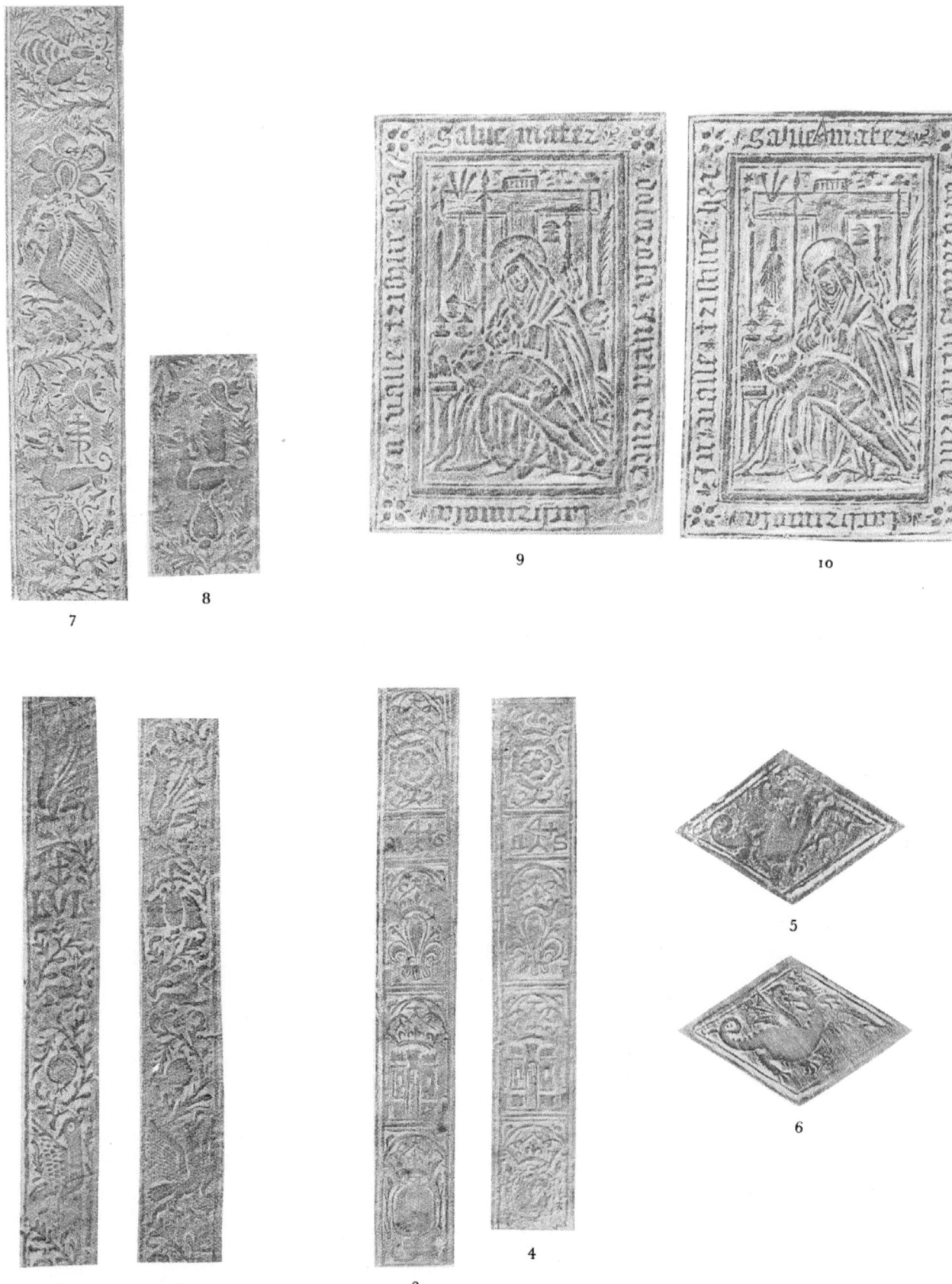

TOOLS IN TWO STATES

PLATE II

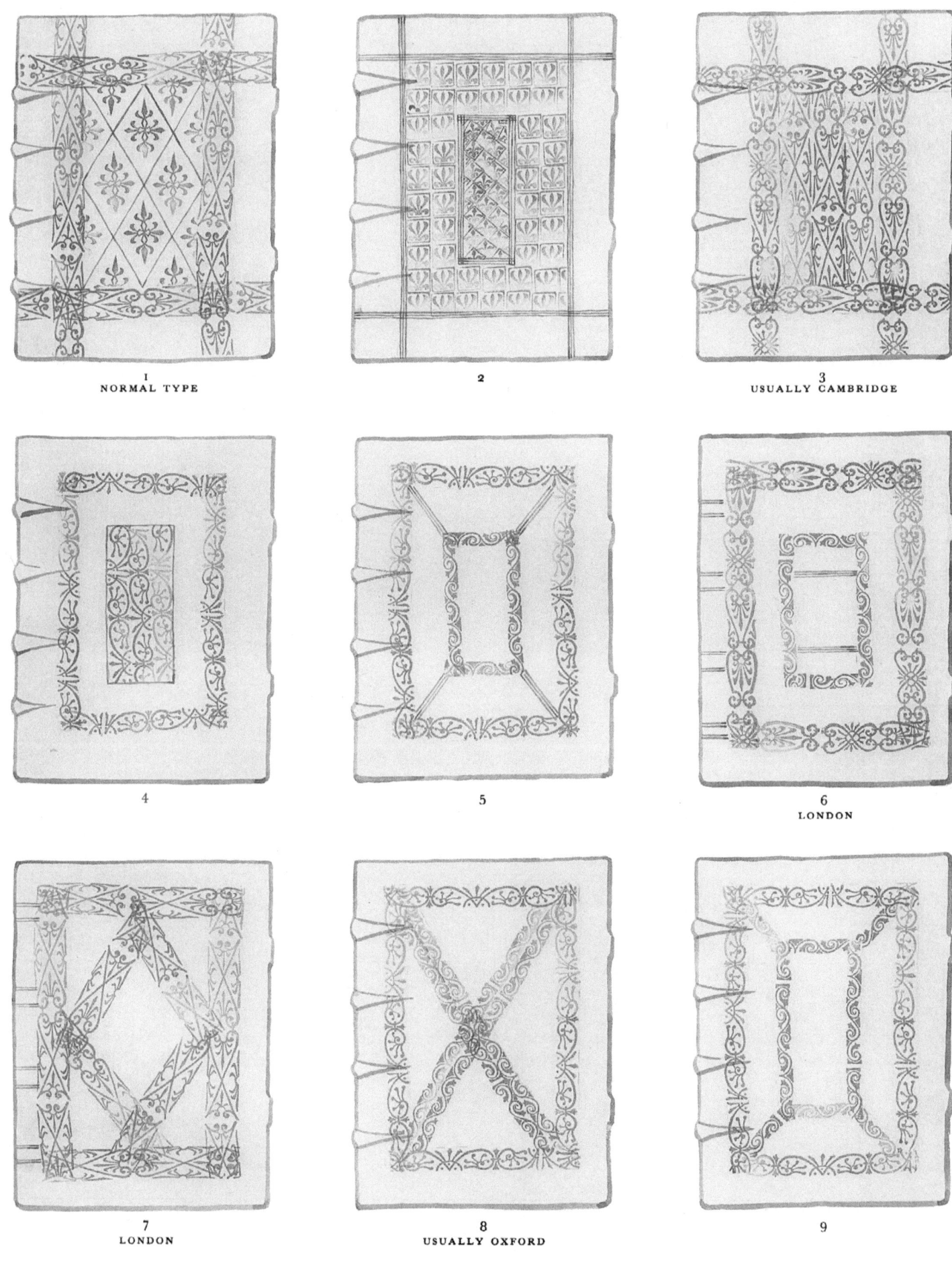

1
NORMAL TYPE

2

3
USUALLY CAMBRIDGE

4

5

6
LONDON

7
LONDON

8
USUALLY OXFORD

9

ENGLISH DESIGNS

PLATE III

PARIS DESIGN
REDUCED

PLATE IV

NETHERLANDISH DESIGN (a)

PLATE V

NETHERLANDISH DESIGN (*b*)
REDUCED

PLATE VI

WOODEN DIE PANEL STAMP
REDUCED

PLATE VI

PLATE VII

GERMAN DESIGN (a)

PLATE VII.

PLATE VIII

GERMAN DESIGN (b)

PLATE IX

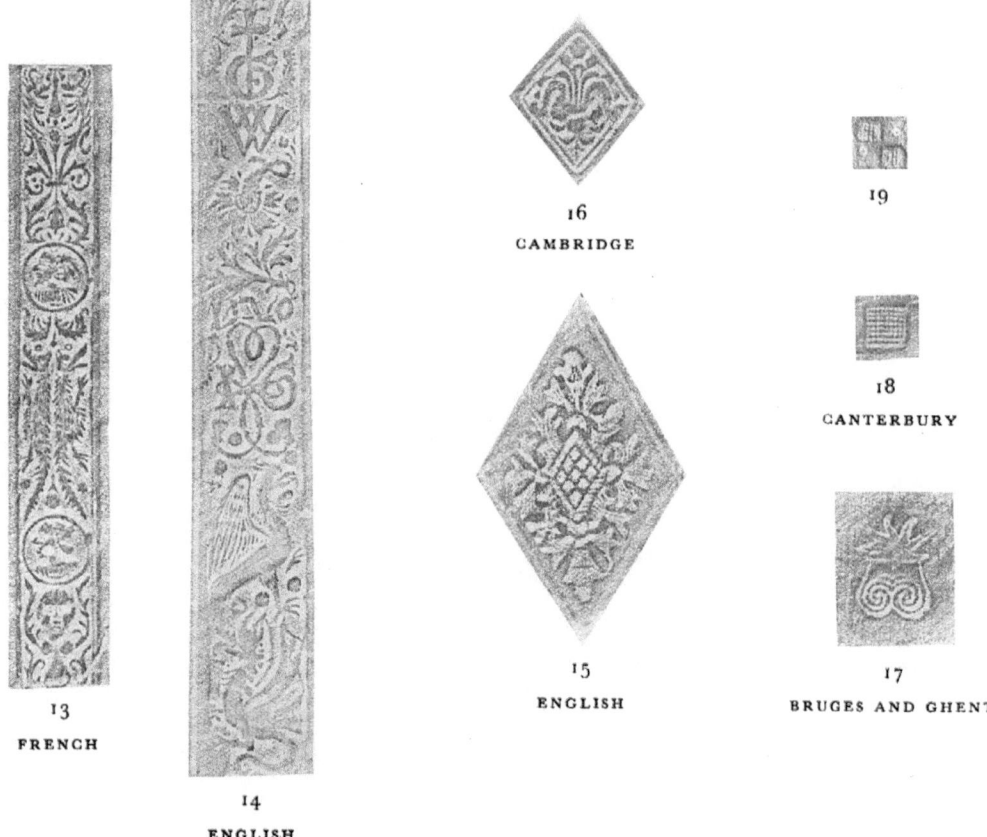

13
FRENCH

14
ENGLISH

16
CAMBRIDGE

15
ENGLISH

19

18
CANTERBURY

17
BRUGES AND GHENT

12
ENGLISH

II
FRENCH

LOCAL TYPES

PLATE X

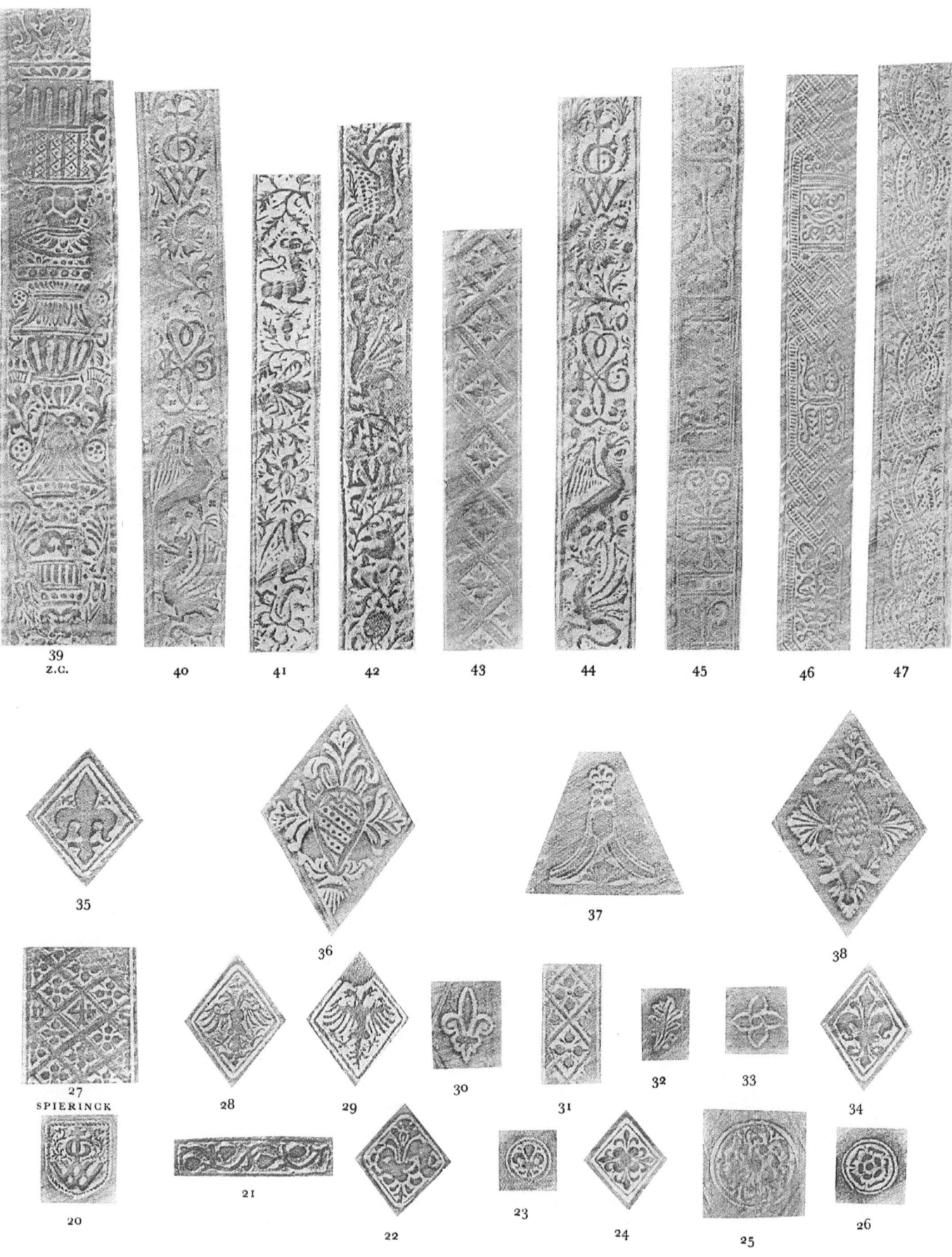

39
Z.C.

40

41

42

43

44

45

46

47

35

36

37

38

27
SPIERINCK

28

29

30

31

32

33

34

20

21

22

23

24

25

26

W.G. AND I.G.

PLATE XI

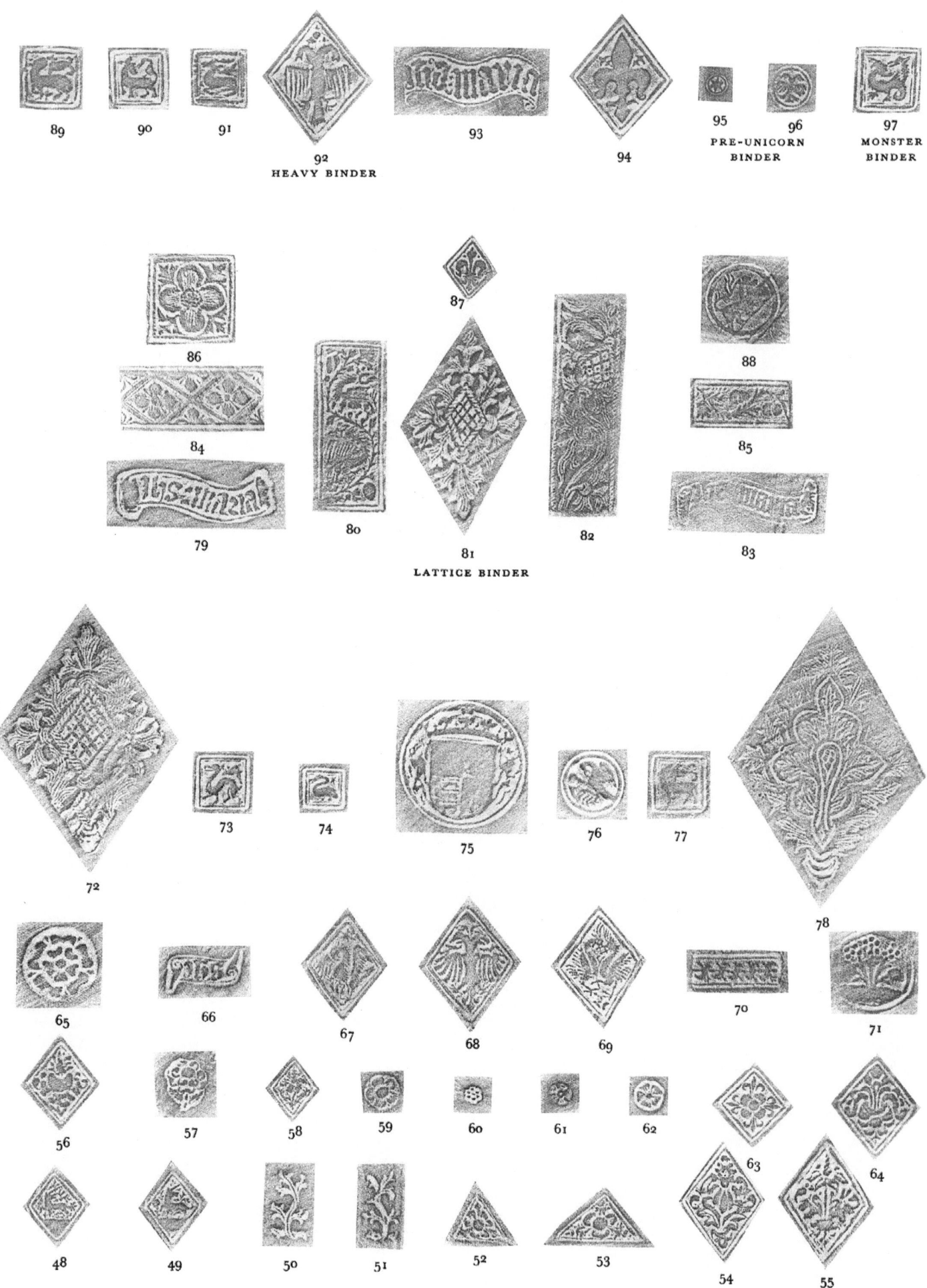

89   90   91

92
HEAVY BINDER

93

94

95
PRE-UNICORN
BINDER

96

97
MONSTER
BINDER

86

87

84

88

79

80

81
LATTICE BINDER

82

83

85

72

73

74

75

76

77

78

65

66

67

68

69

70

71

56

57

58

59

60

61

62

63

64

48

49

50

51

52

53

54

55

UNICORN BINDER

PLATE XII

ANTWERP BINDER
REDUCED

PLATE XIII

ATHOS BINDER
REDUCED

PLATE XII.

PLATE XIV

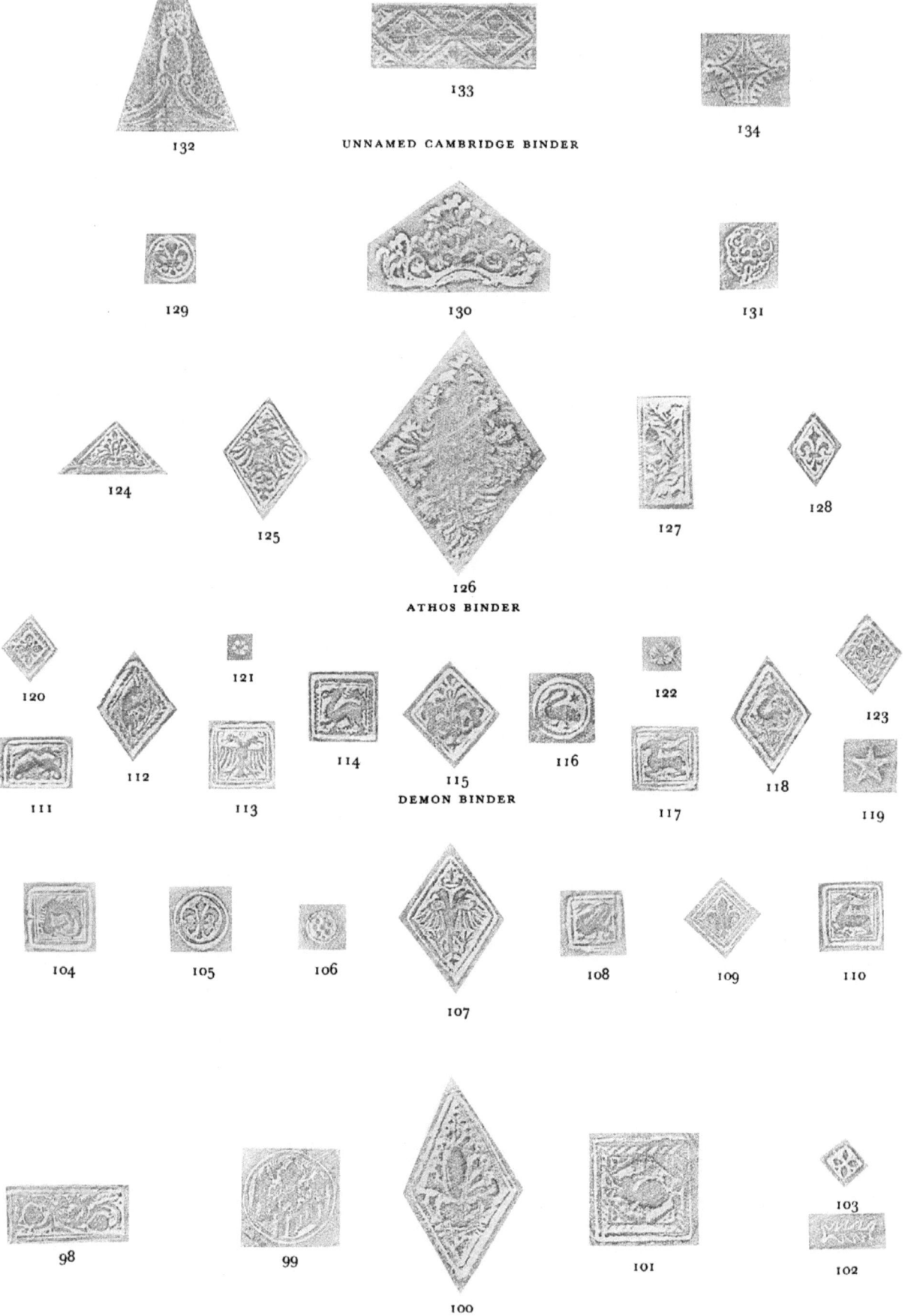

132

133

UNNAMED CAMBRIDGE BINDER

134

129

130

131

124

125

126

ATHOS BINDER

127

128

120

121

122

123

112

114

115

116

118

111

113

DEMON BINDER

117

119

104

105

106

107

108

109

110

98

99

100

101

102

103

ANTWERP BINDER

PLATE XV

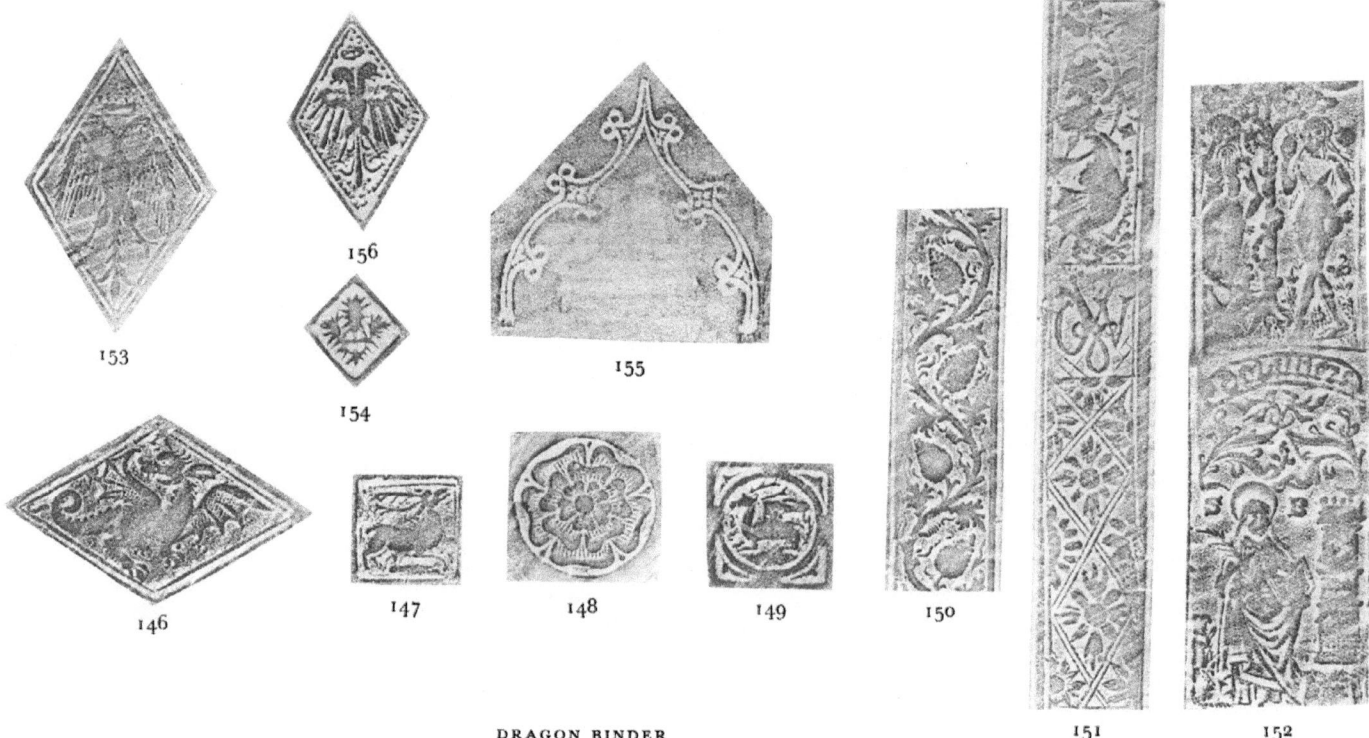

153   156   154   155   146   147   148   149   150   151   152

DRAGON BINDER

145

140   141   142   143   144   135   136   137   138   139

GREYHOUND BINDER

PLATE XVI

FISHTAIL BINDER
REDUCED

PLATE XVII

FRUIT AND FLOWER BINDER
REDUCED

PLATE XVII

PLATE XVIII

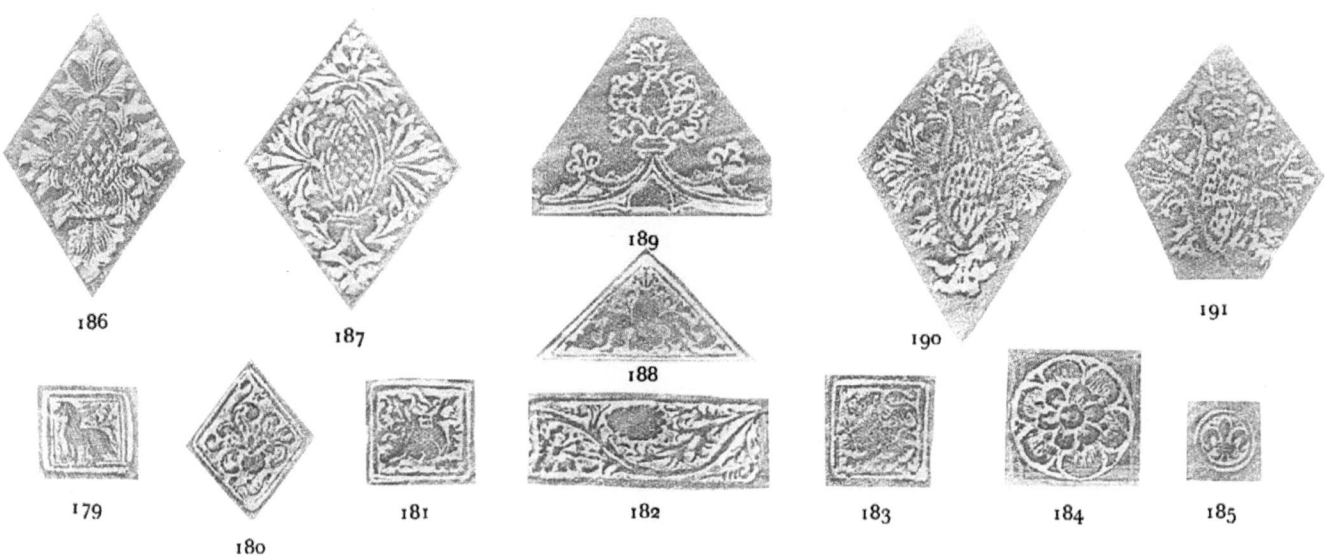

FRUIT AND FLOWER BINDER

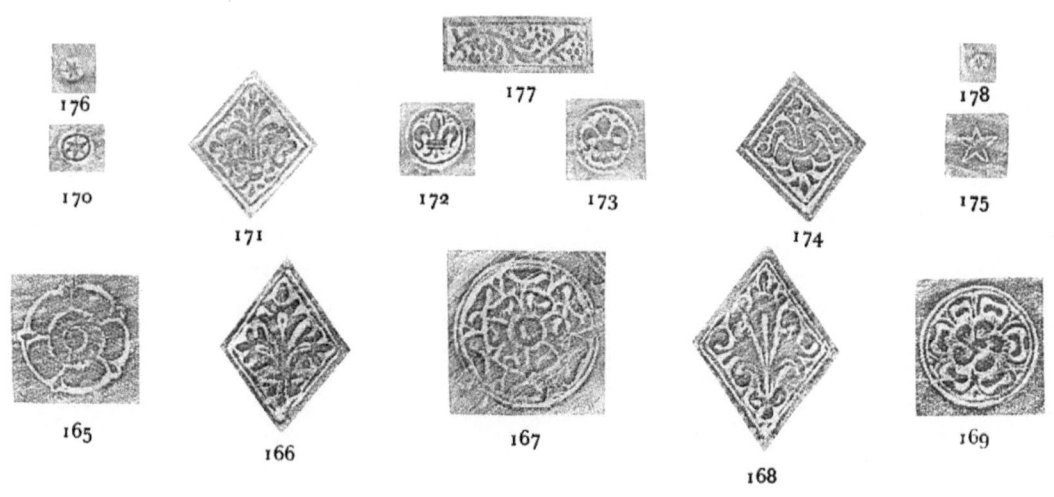

FLORAL BINDER

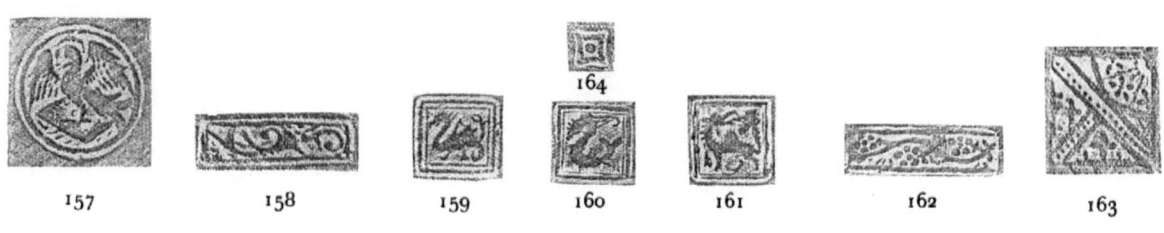

FISHTAIL BINDER

PLATE XIX

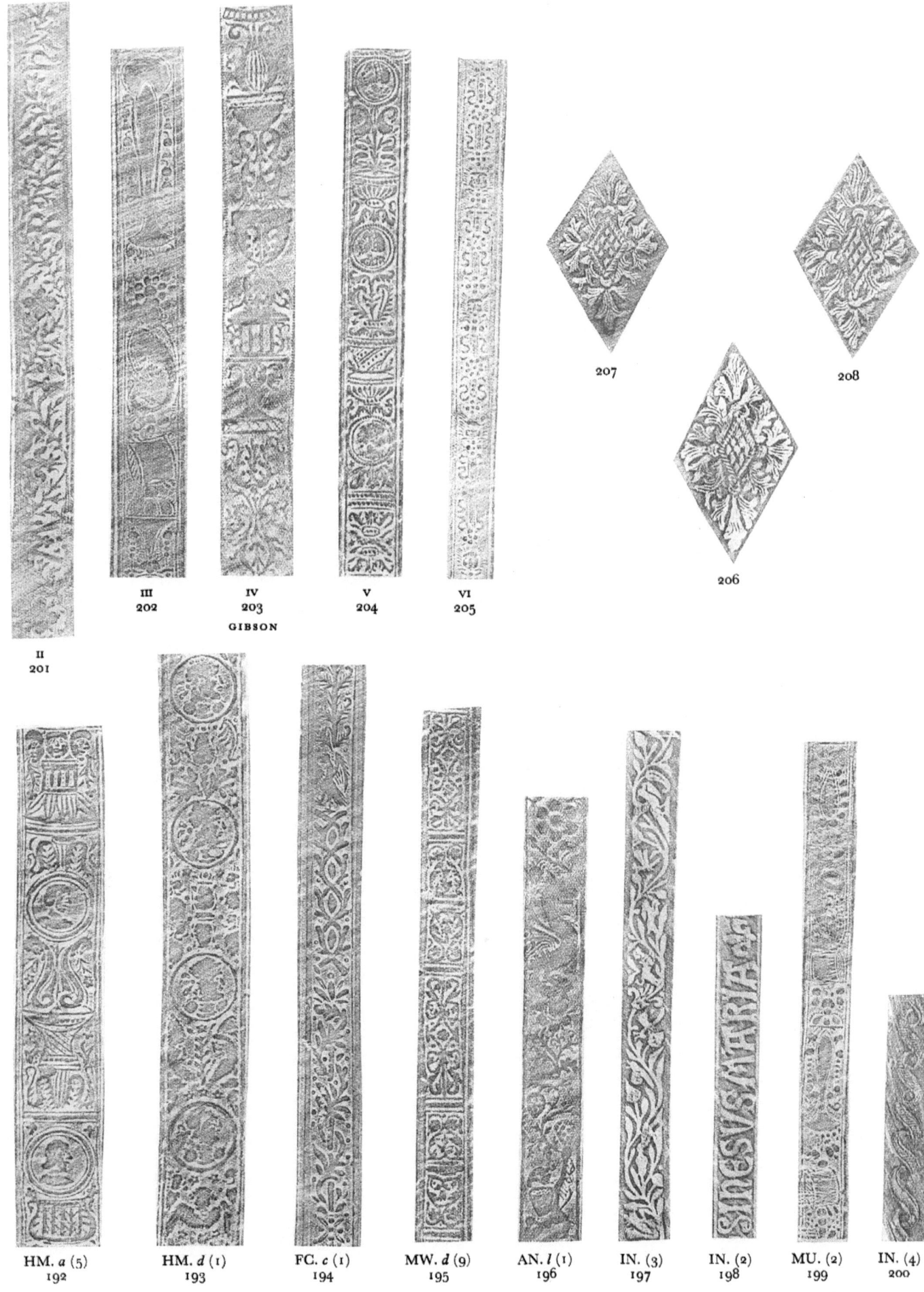

III
202

IV
203

V
204

VI
205

GIBSON

II
201

207

208

206

HM. *a* (5)
192

HM. *d* (1)
193

FC. *c* (1)
194

MW. *d* (9)
195

AN. *l* (1)
196

IN. (3)
197

IN. (2)
198

MU. (2)
199

IN. (4)
200

PLATE XX

232

225   226   227   228   229   230   231

216   217   218   219   220   221   222   223   224

209   210   211   212   213   214   215

CANTERBURY

PLATE XXI

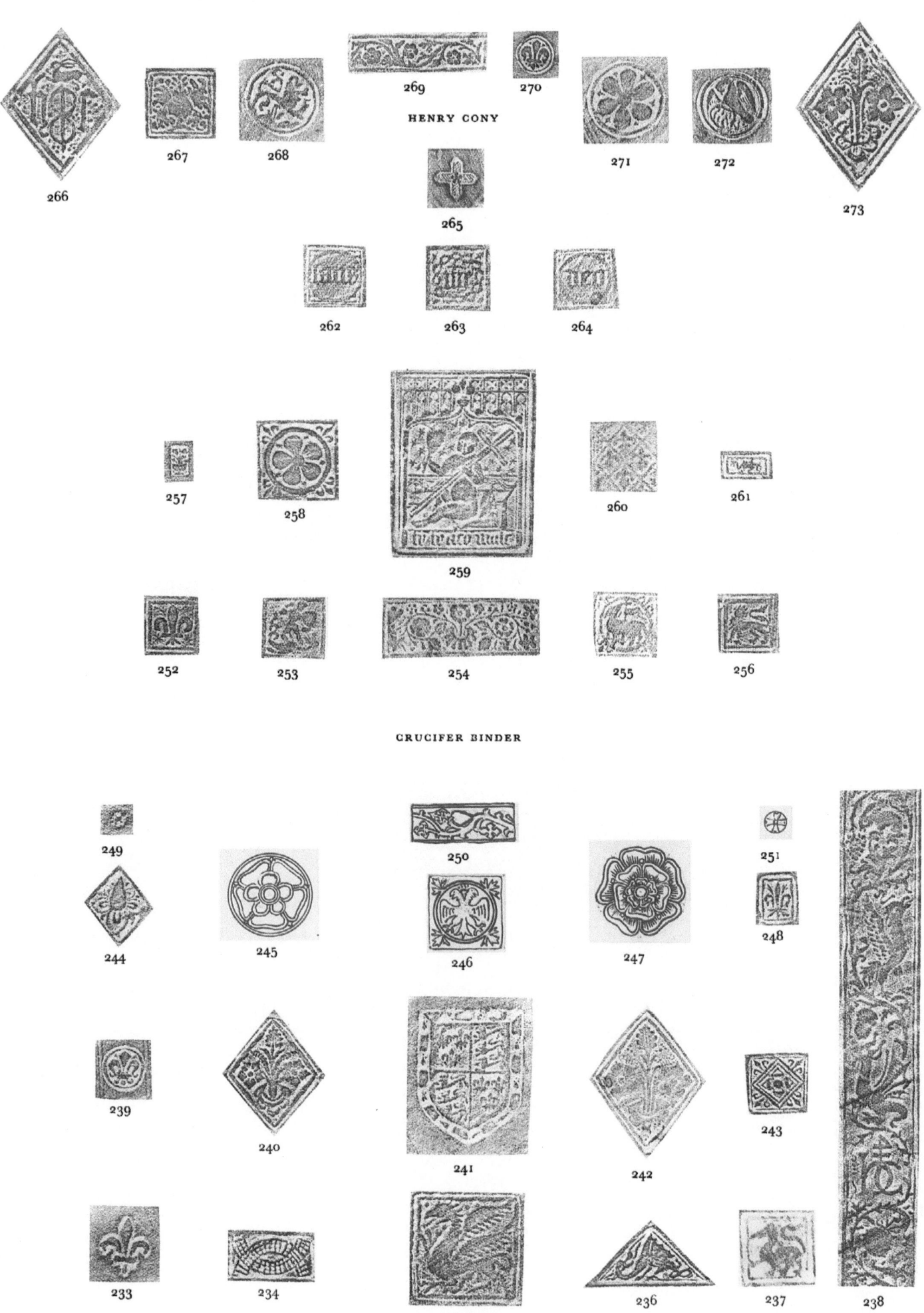

HENRY CONY

CRUCIFER BINDER

CAXTON BINDERY

PLATE XXII

BAT BINDER

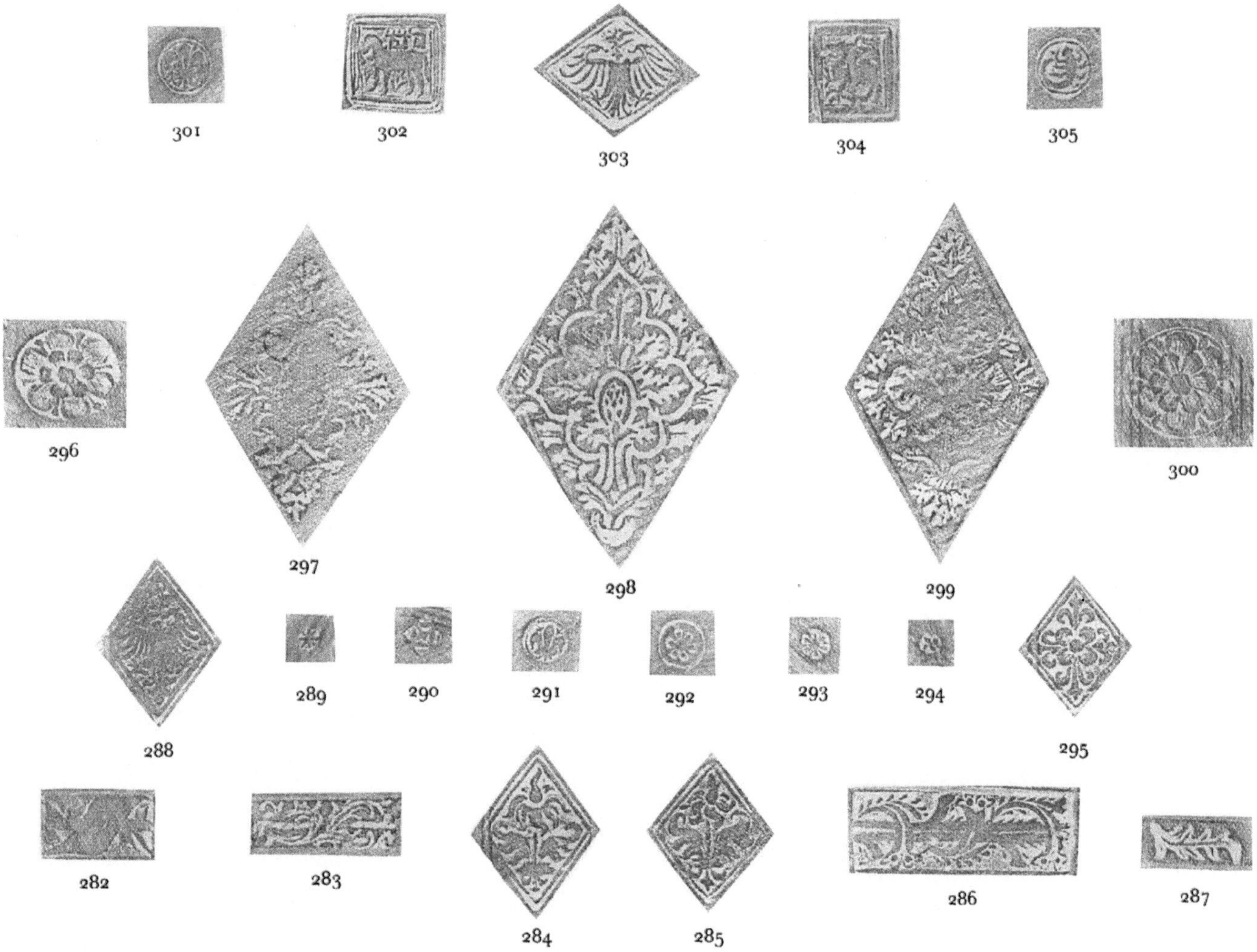

PLATE XXIII

301
302
303
304
305

296
297
298
299
300

288
289
290
291
292
293
294
295

282
283
284
285
286
287

LILY BINDER

279
280
281

278

275
276
277

274

BAT BINDER

PLATE XXIII

PLATE XXIV

HALF-STAMP BINDER

PLATE XXV

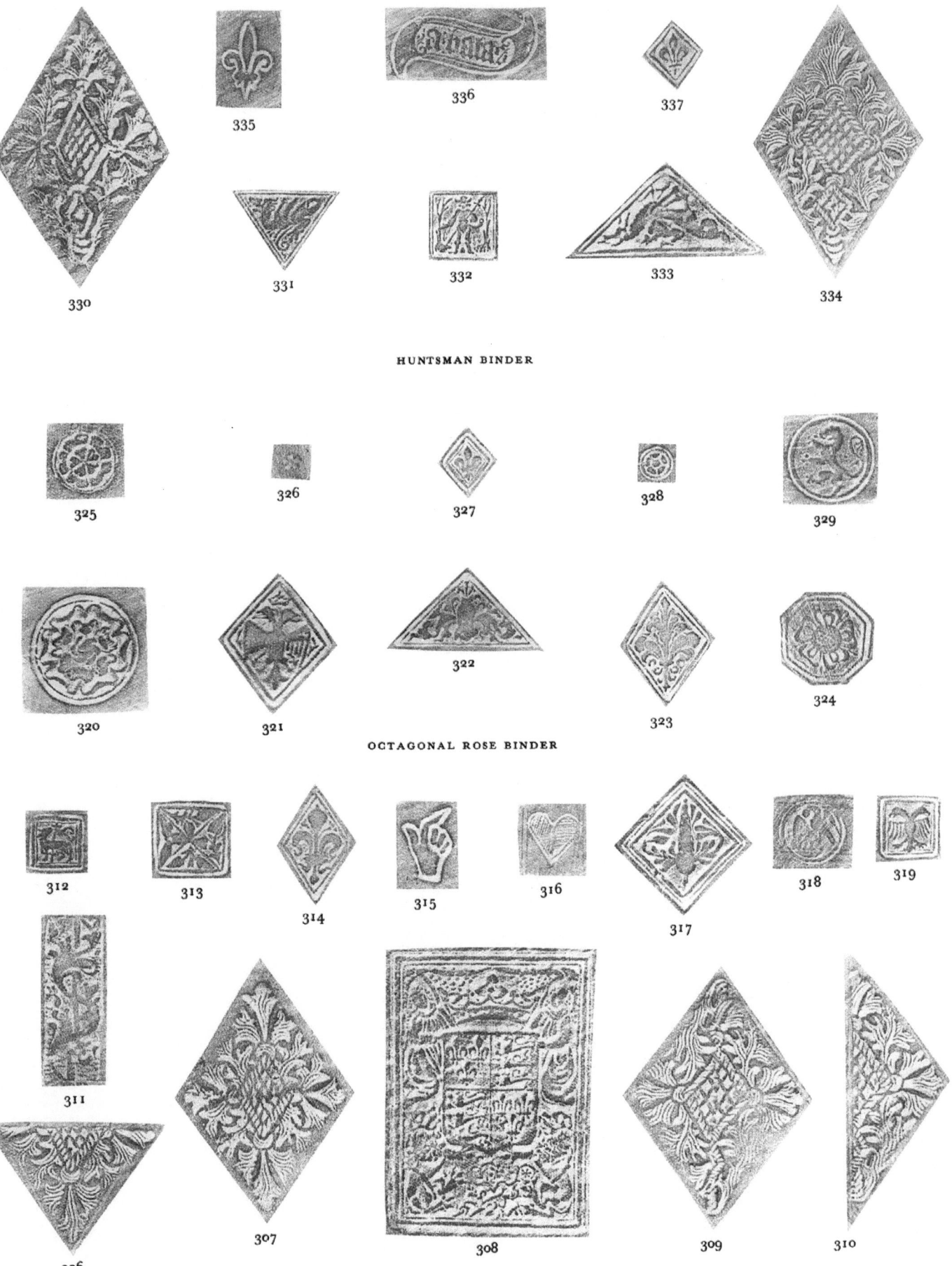

335

336

337

330

331

332

333

334

HUNTSMAN BINDER

325

326

327

328

329

320

321

322

323

324

OCTAGONAL ROSE BINDER

312

313

314

315

316

317

318

319

311

306

307

308

309

310

HALF-STAMP BINDER

PLATE XXVI

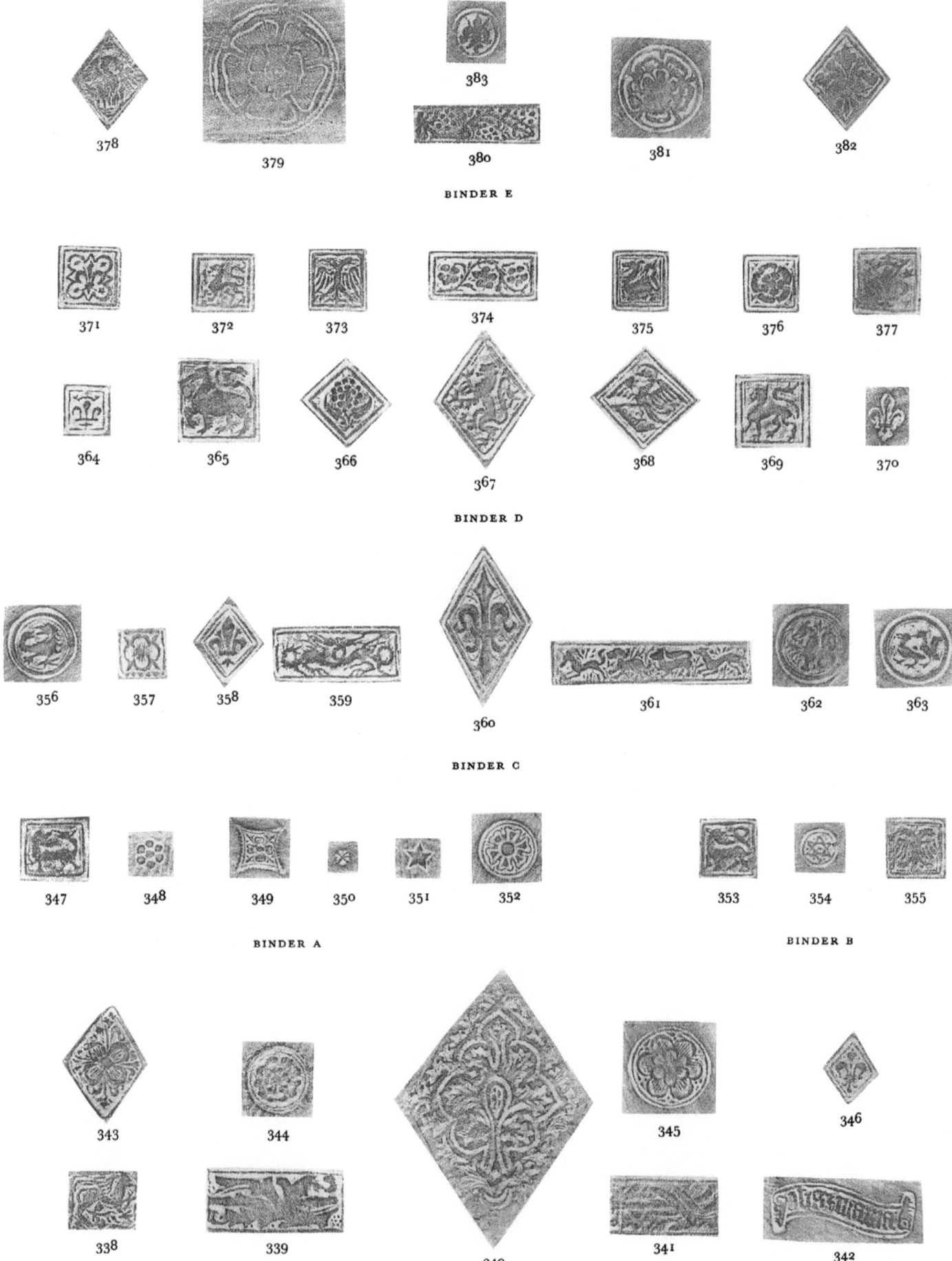

378
379
383
380
381
382

BINDER E

371 372 373 374 375 376 377

364 365 366 367 368 369 370

BINDER D

356 357 358 359 360 361 362 363

BINDER C

347 348 349 350 351 352 353 354 355

BINDER A   BINDER B

343 344 340 345 346

338 339 341 342

FOLIAGED STAFF BINDER

PLATE XXVII

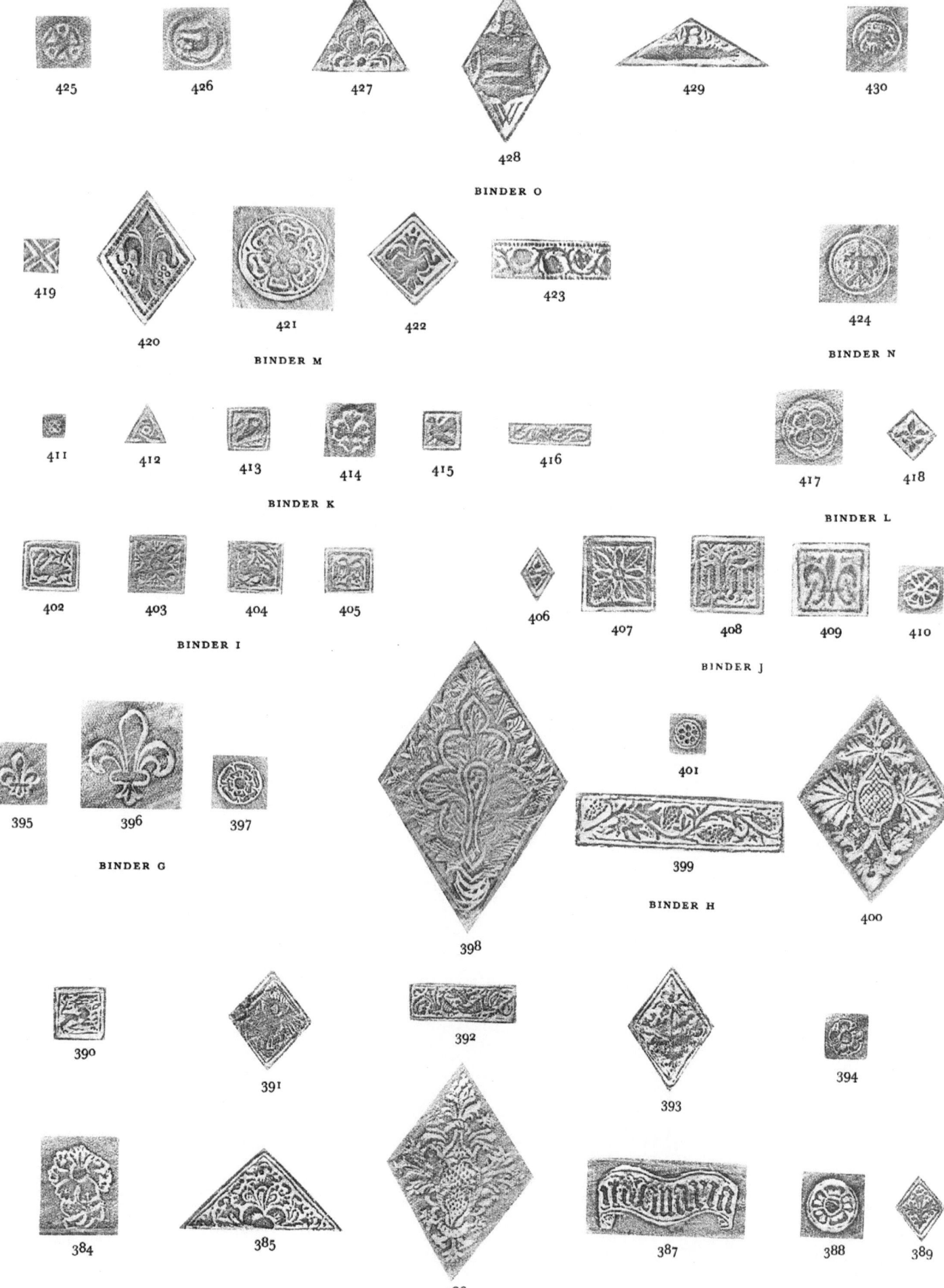

425

426

427

428

429

430

BINDER O

419

420

421

422

423

424

BINDER M

BINDER N

411

412

413

414

415

416

417

418

BINDER K

BINDER L

402

403

404

405

406

407

408

409

410

BINDER I

BINDER J

395

396

397

BINDER G

398

399

400

401

BINDER H

390

391

392

393

394

384

385

386

387

388

389

BINDER F

PLATE XXVIII

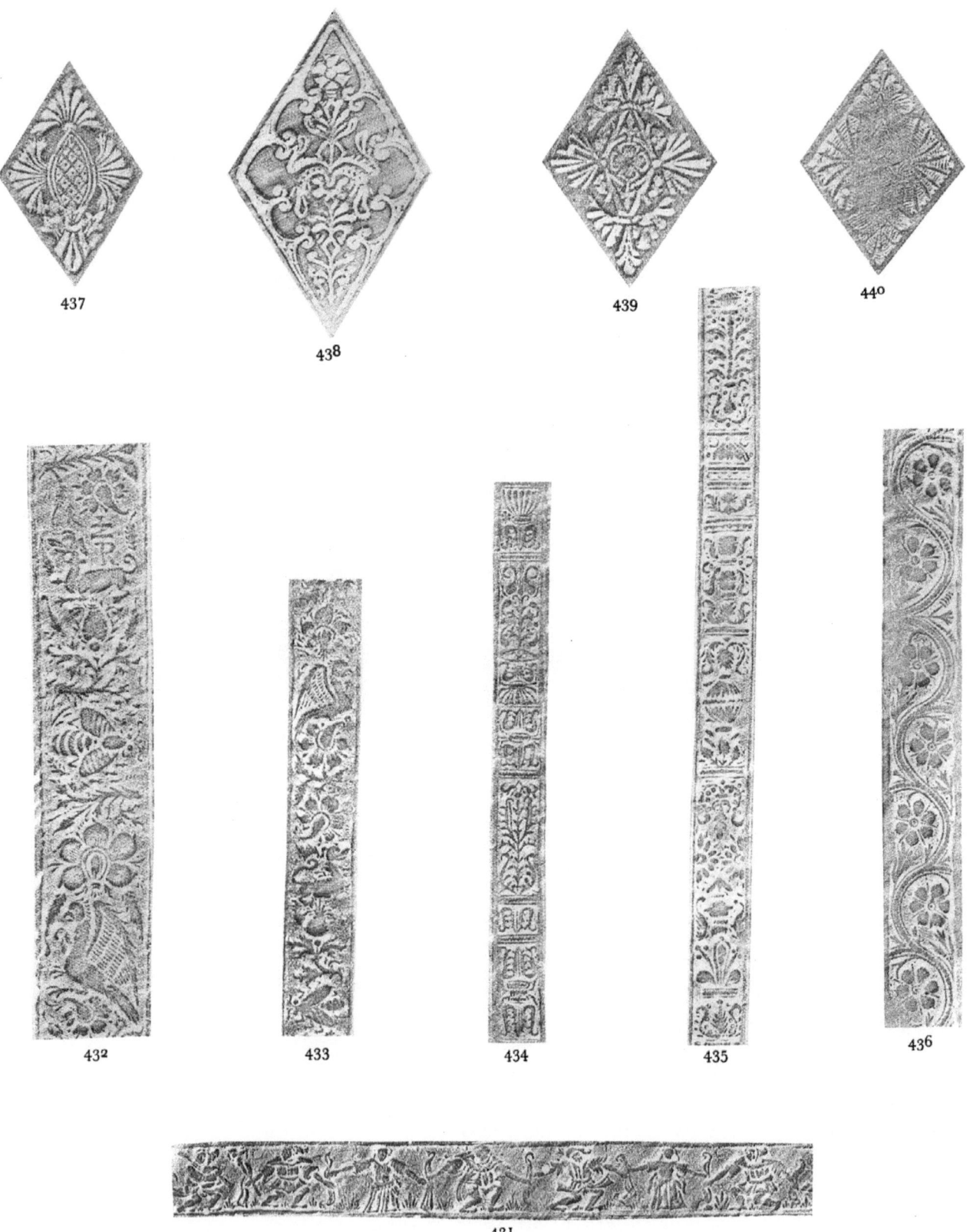

437

438

439

440

432

433

434

435

436

431

REYNES

PLATE XXIX

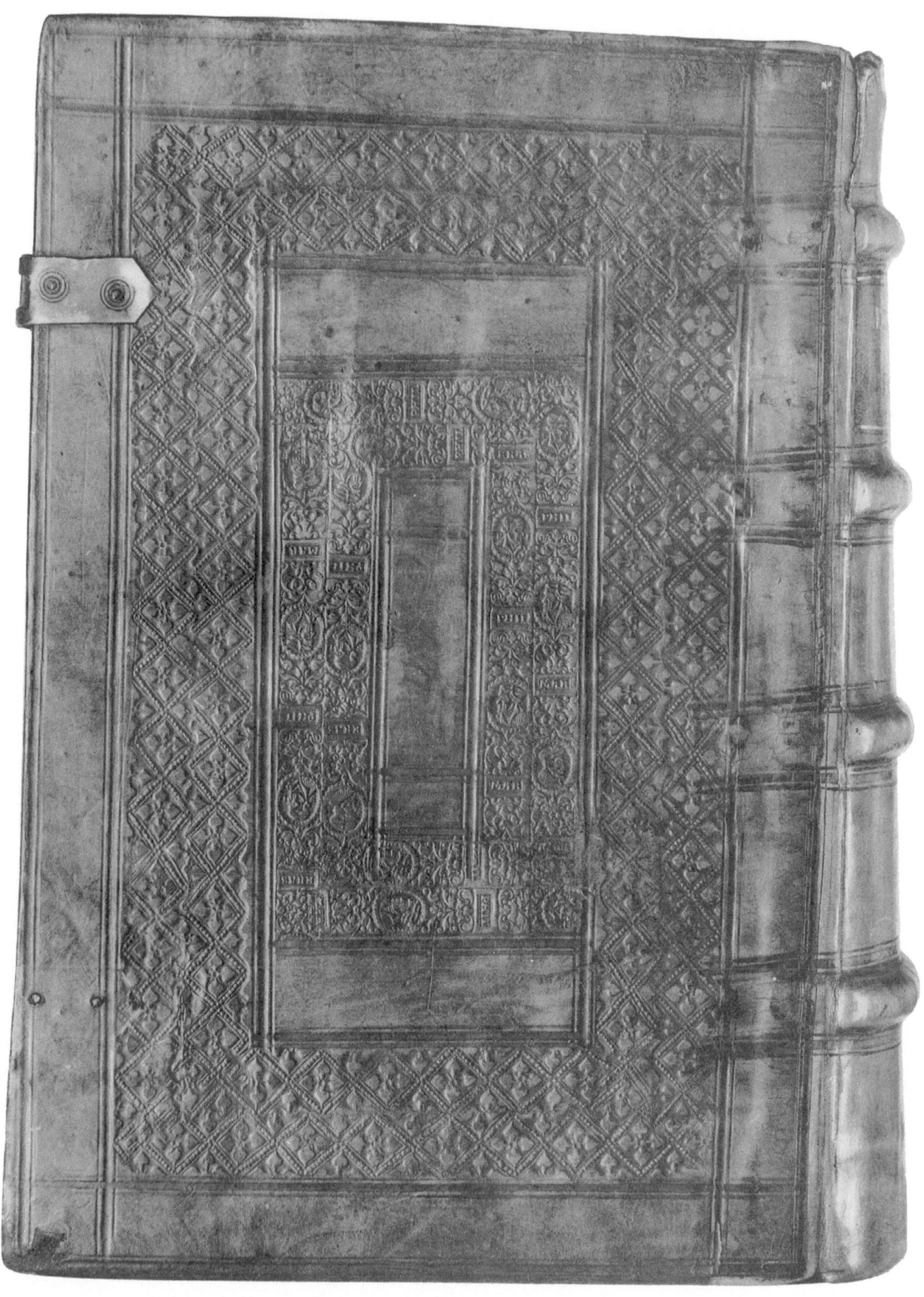

BINDER F.D.

PLATE XXX

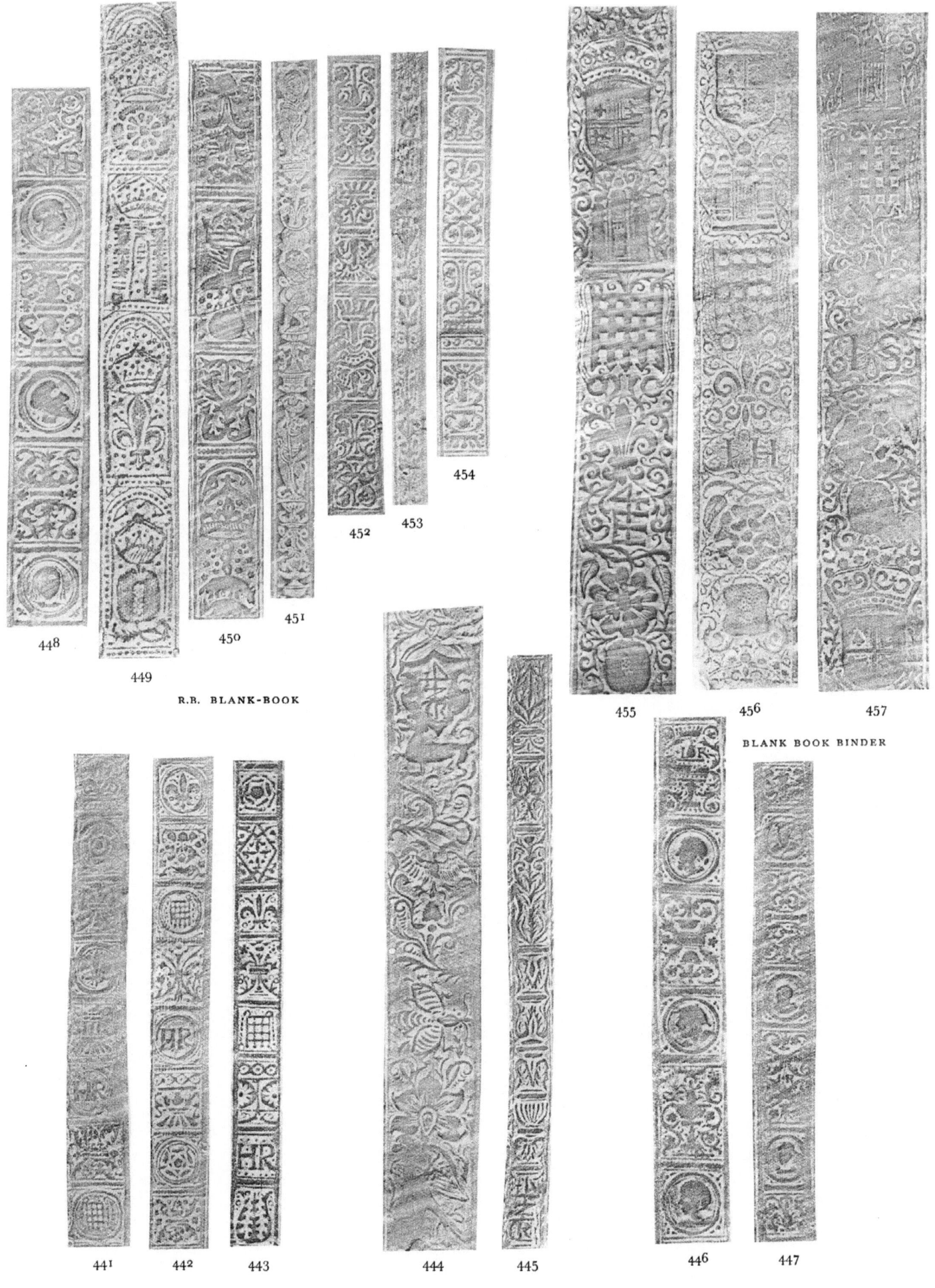

448

449

R.B. BLANK-BOOK

450

451

452

453

454

455   456   457

BLANK BOOK BINDER

441   442   443   444   445   446   447

H.R. AND I.R.

PLATE XXXI

BLANK-BOOK BINDER
REDUCED

PLATE XXXII

475

476

477

478

467

468

473

474

470

471

472

469

458

459

460

461

465

466

462

463

464

POSSIBLE DERIVATIONS OF ORNAMENT

PLATE XXXIII

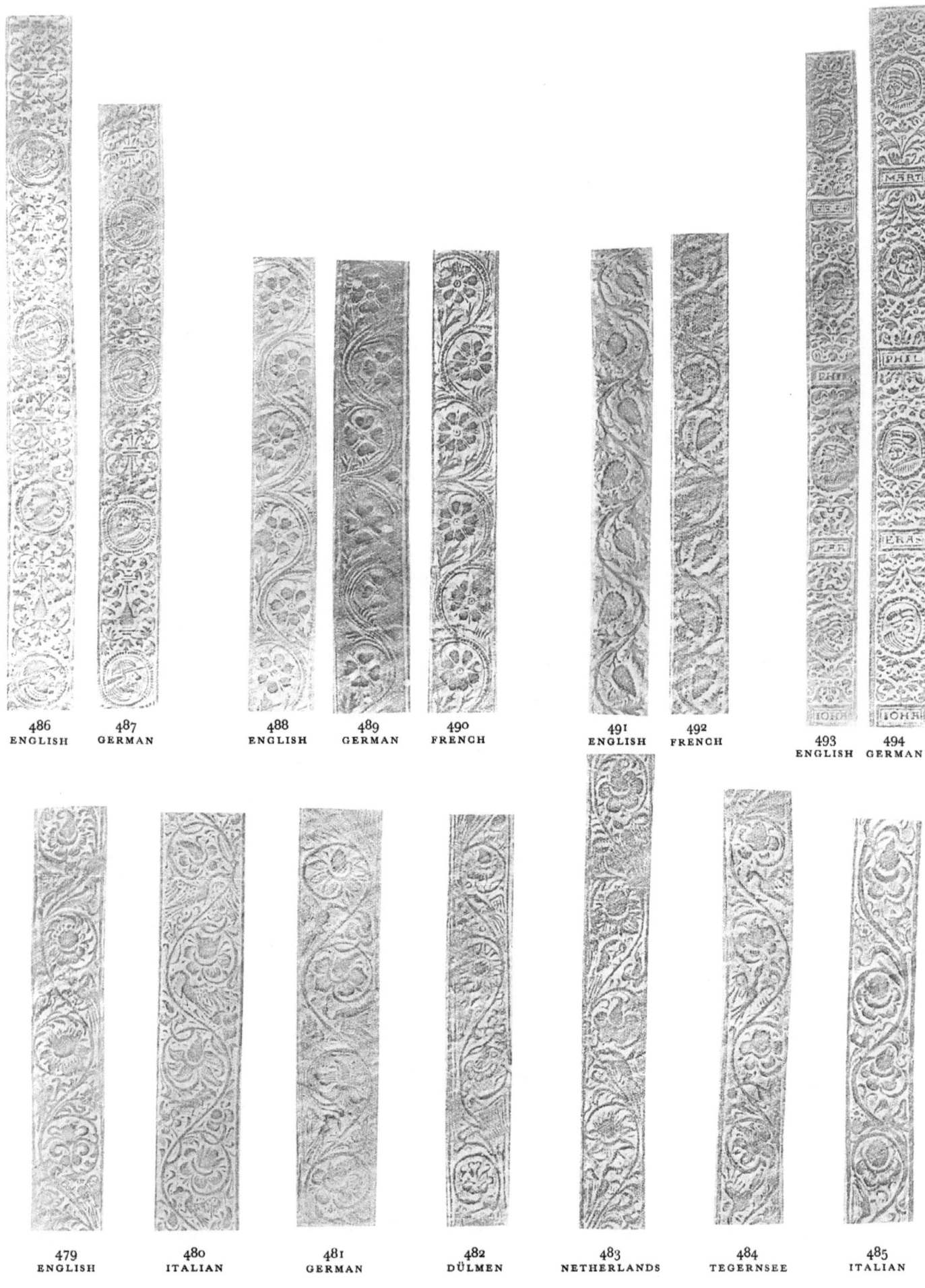

486
ENGLISH

487
GERMAN

488
ENGLISH

489
GERMAN

490
FRENCH

491
ENGLISH

492
FRENCH

493
ENGLISH

494
GERMAN

479
ENGLISH

480
ITALIAN

481
GERMAN

482
DÜLMEN

483
NETHERLANDS

484
TEGERNSEE

485
ITALIAN

ENGLISH AND FOREIGN VARIANTS

PLATE XXIII

Fig. 1. A. FORMIO REGIANS

PLATE XXXIV

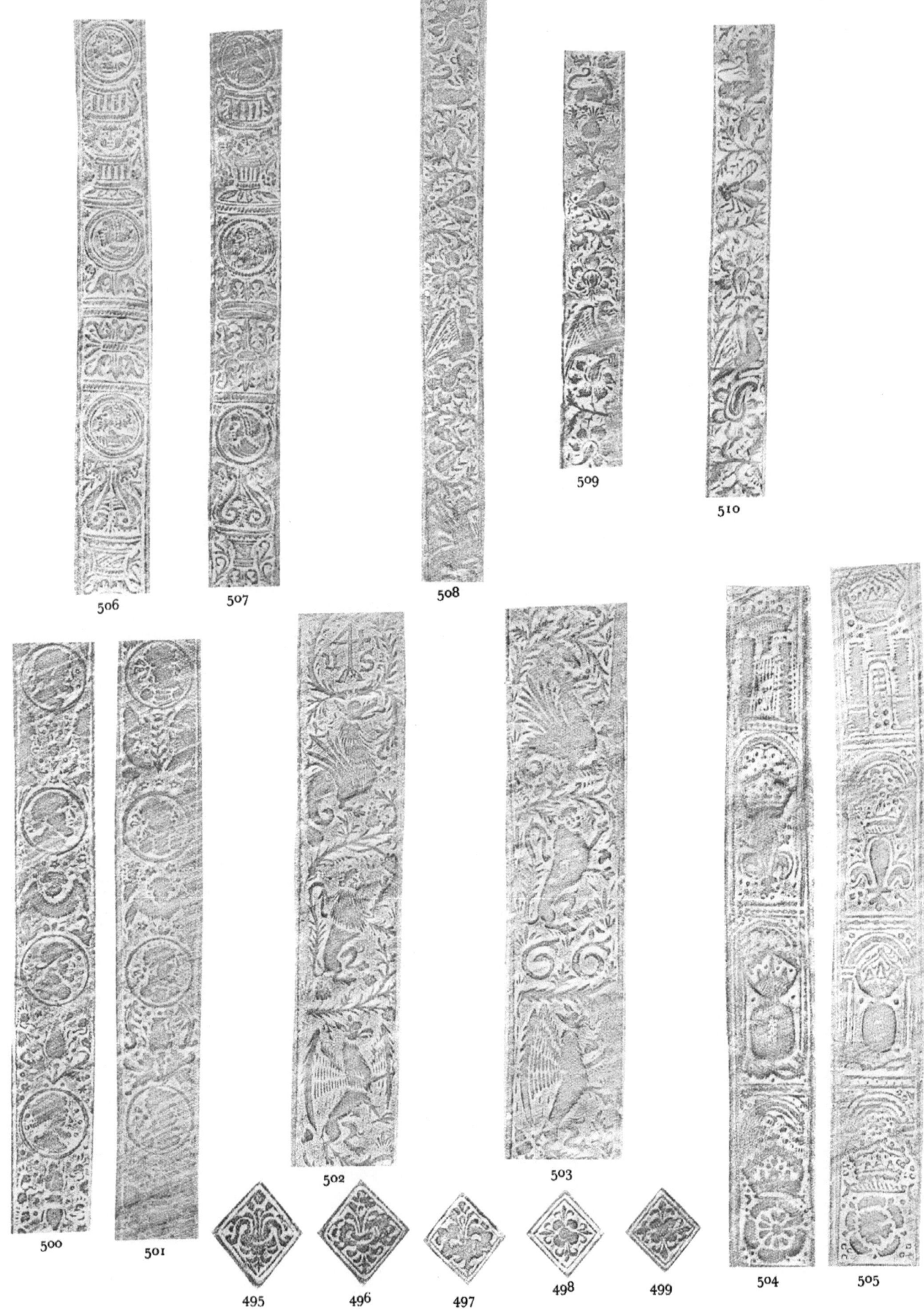

506 507 508 509 510

500 501 502 503 504 505

495 496 497 498 499

ENGLISH VARIANTS

PLATE XXXV

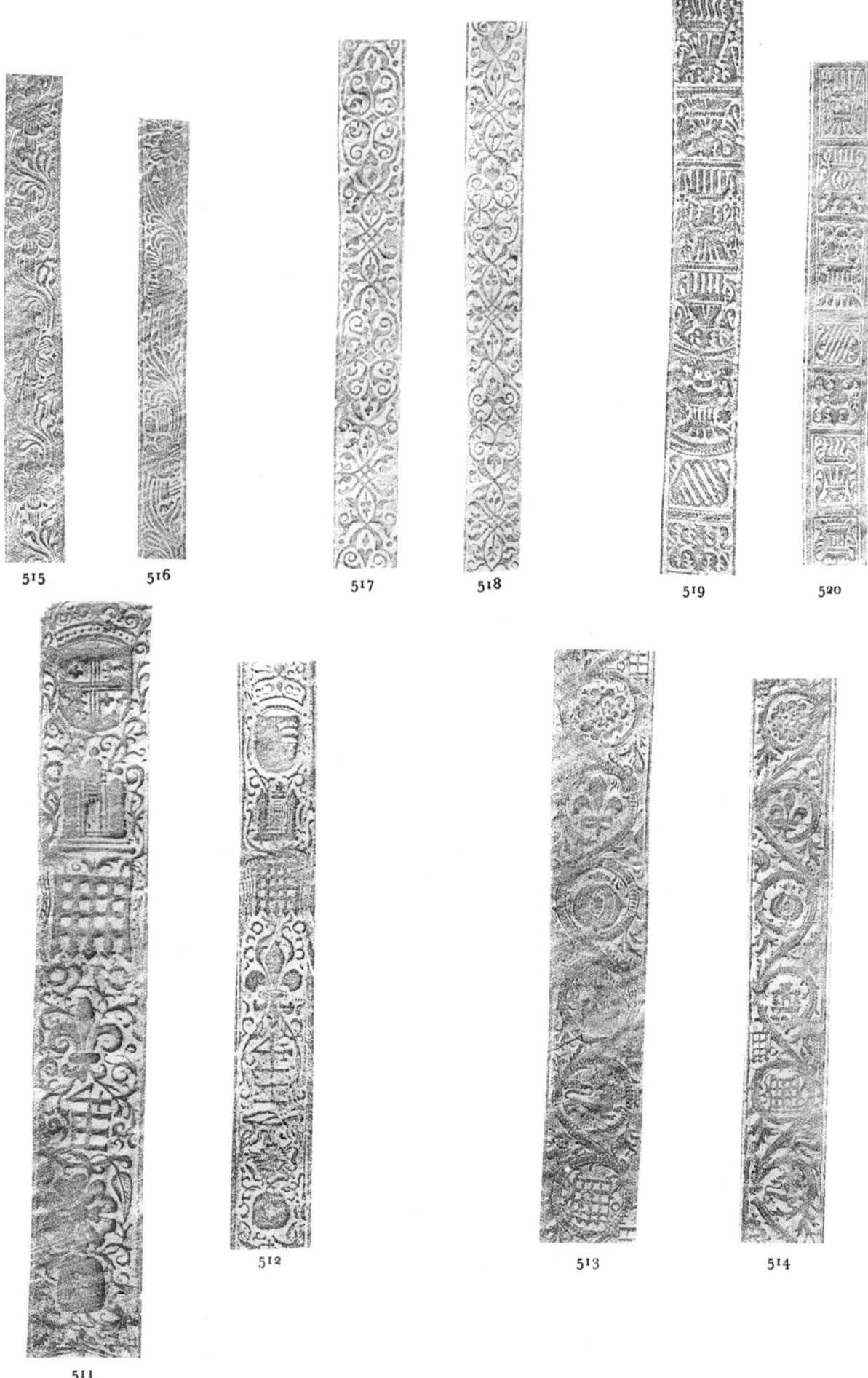

515   516   517   518   519   520

511   512   513   514

VARIANTS IN DIFFERENT SIZES

PLATE XXXVI

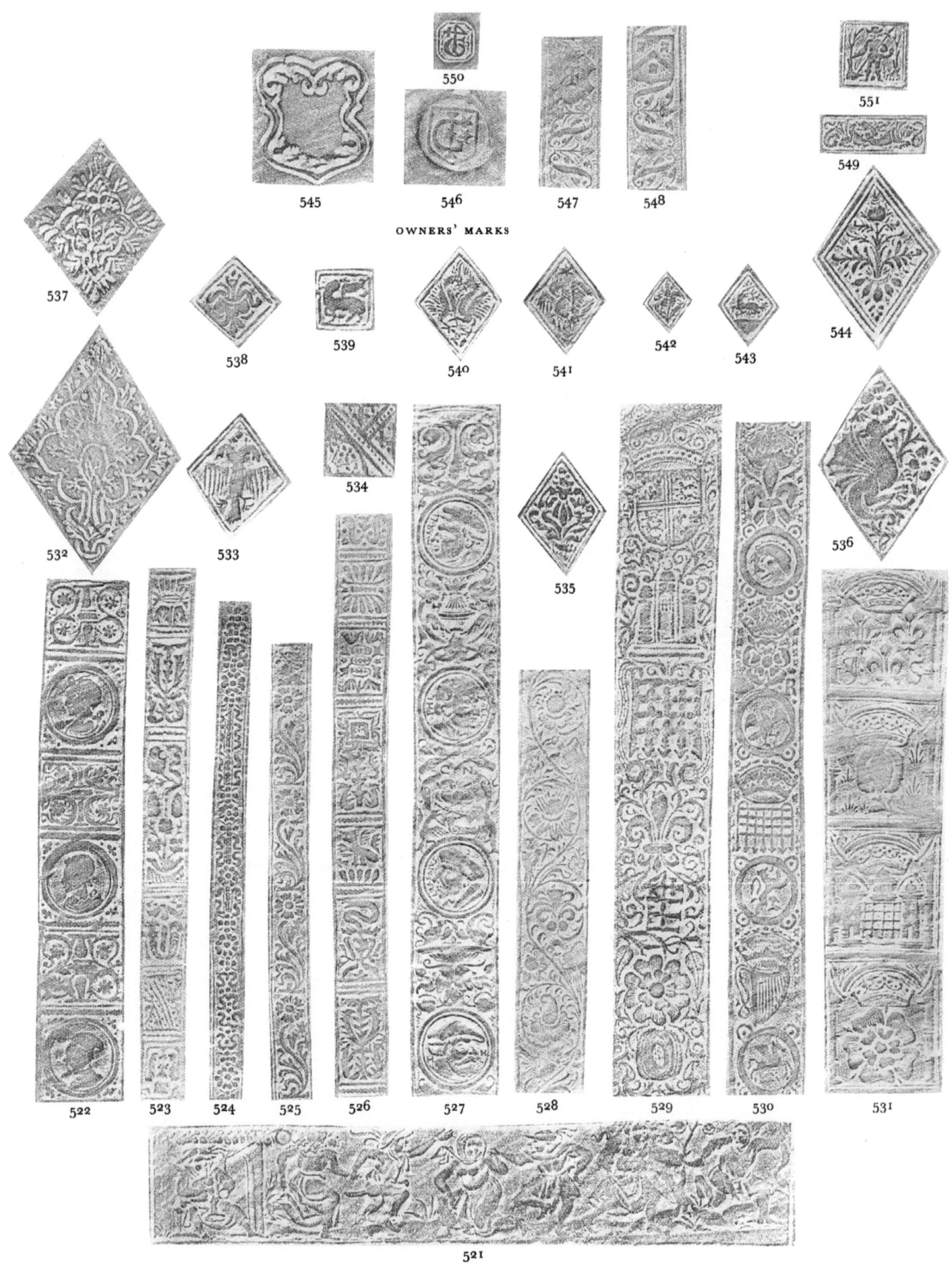

OWNERS' MARKS

HOLBEIN

BAD AND GOOD DESIGNS

PLATE XXXVII

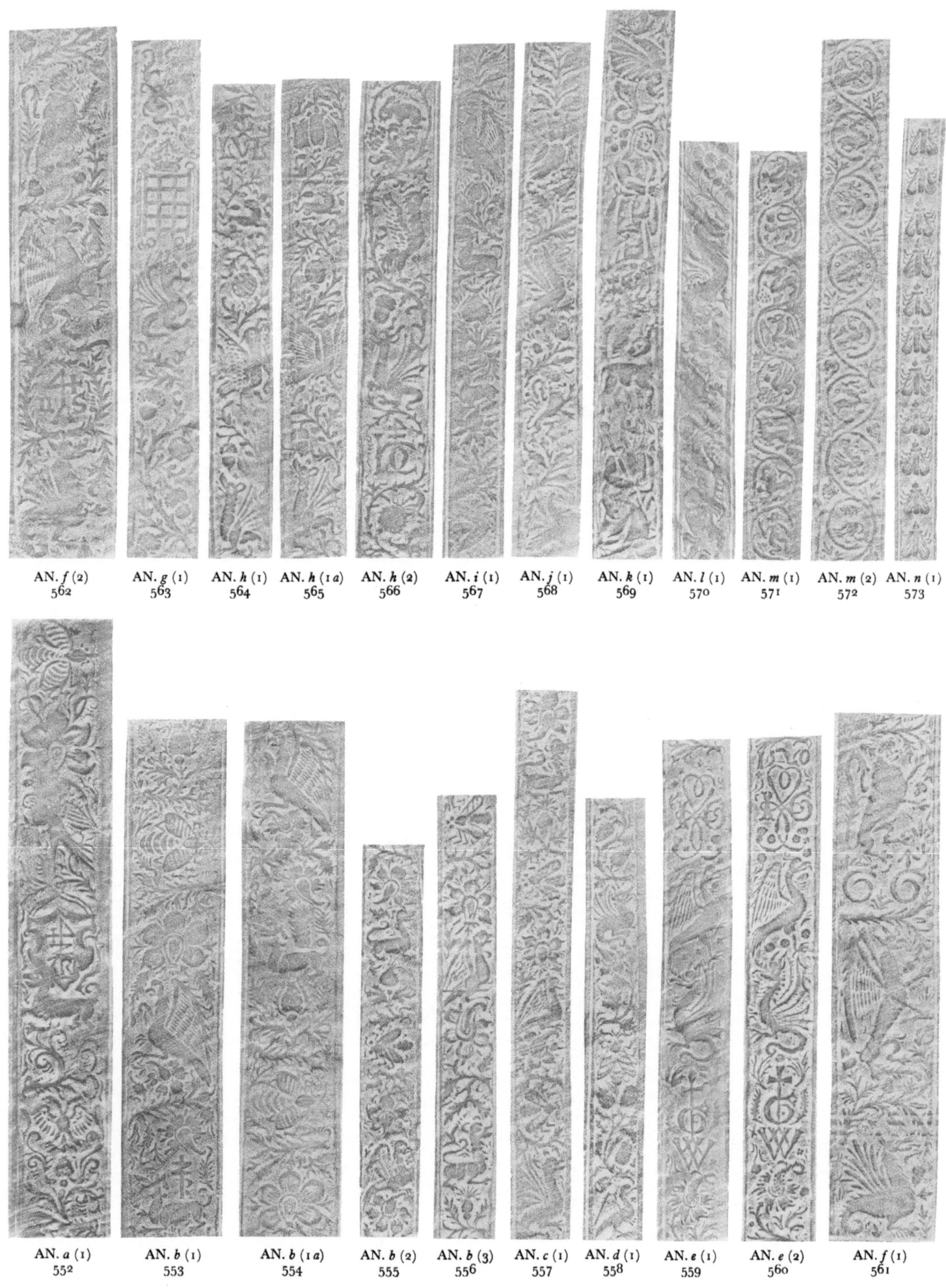

AN. *f* (2)
562

AN. *g* (1)
563

AN. *h* (1)
564

AN. *h* (1 *a*)
565

AN. *h* (2)
566

AN. *i* (1)
567

AN. *j* (1)
568

AN. *k* (1)
569

AN. *l* (1)
570

AN. *m* (1)
571

AN. *m* (2)
572

AN. *n* (1)
573

AN. *a* (1)
552

AN. *b* (1)
553

AN. *b* (1 *a*)
554

AN. *b* (2)
555

AN. *b* (3)
556

AN. *c* (1)
557

AN. *d* (1)
558

AN. *e* (1)
559

AN. *e* (2)
560

AN. *f* (1)
561

PLATE XXXVII.

PLATE XXXVIII

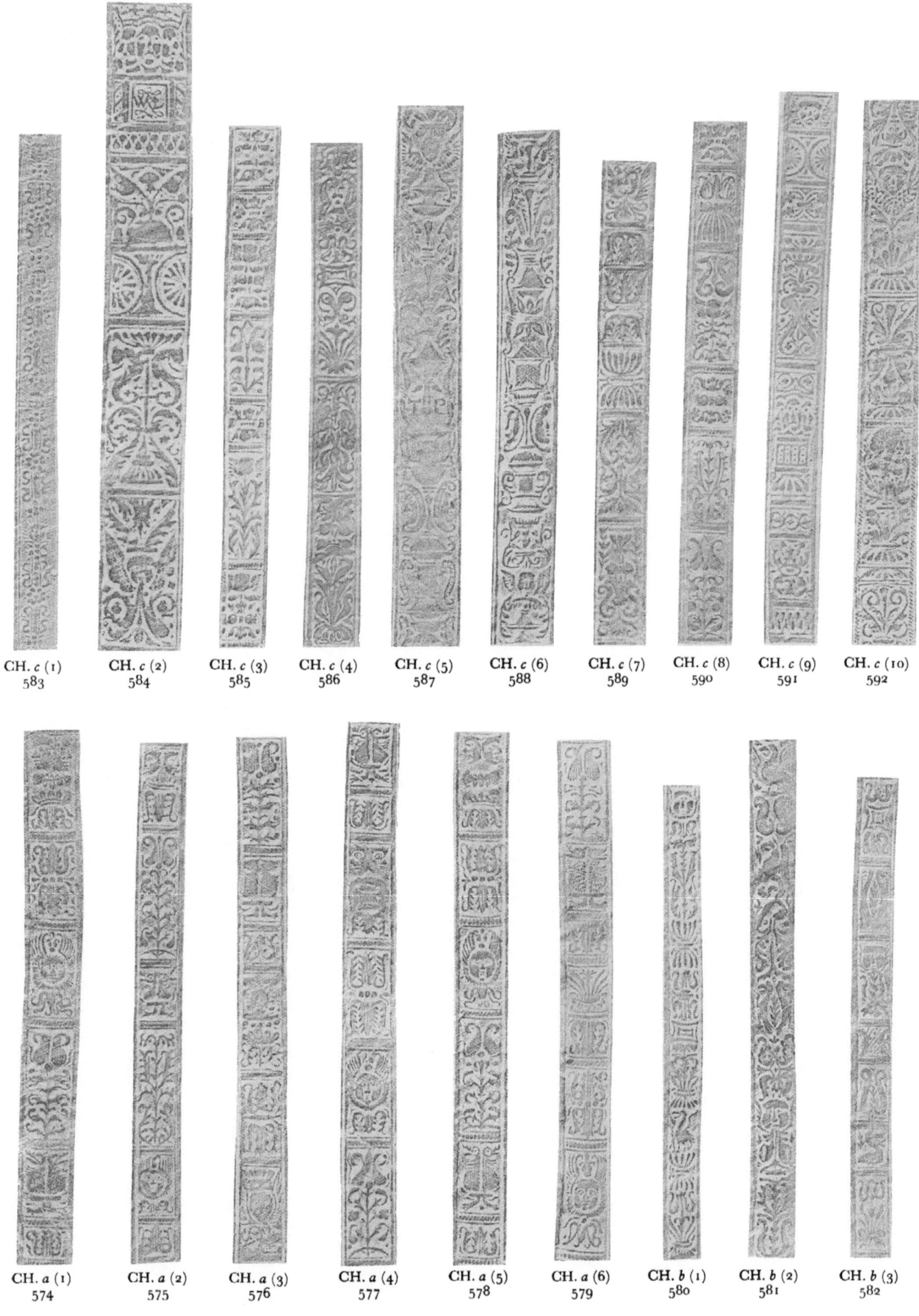

CH. *c* (1)
583

CH. *c* (2)
584

CH. *c* (3)
585

CH. *c* (4)
586

CH. *c* (5)
587

CH. *c* (6)
588

CH. *c* (7)
589

CH. *c* (8)
590

CH. *c* (9)
591

CH. *c* (10)
592

CH. *a* (1)
574

CH. *a* (2)
575

CH. *a* (3)
576

CH. *a* (4)
577

CH. *a* (5)
578

CH. *a* (6)
579

CH. *b* (1)
580

CH. *b* (2)
581

CH. *b* (3)
582

CLASSIFIED ROLLS CH.

PLATE XXXVII

PLATE XXXIX

DI. *b* (1)
606

DI. *c* (1)
607

DI. *d* (1)
608

DI. *d* (2)
609

DI. *e* (1)
610

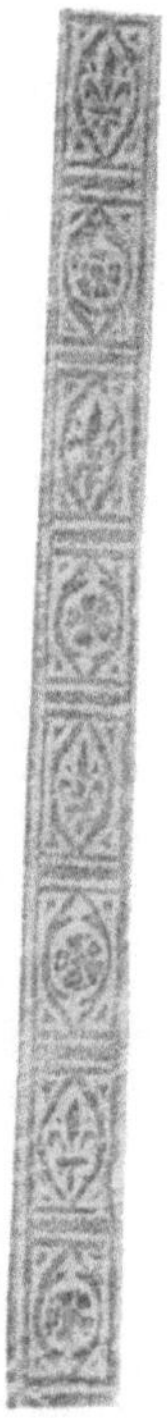

DI. *e* (2)
611

DI. *e* (3)
612

DI. *f* (1)
613

DI. *g* (1)
614

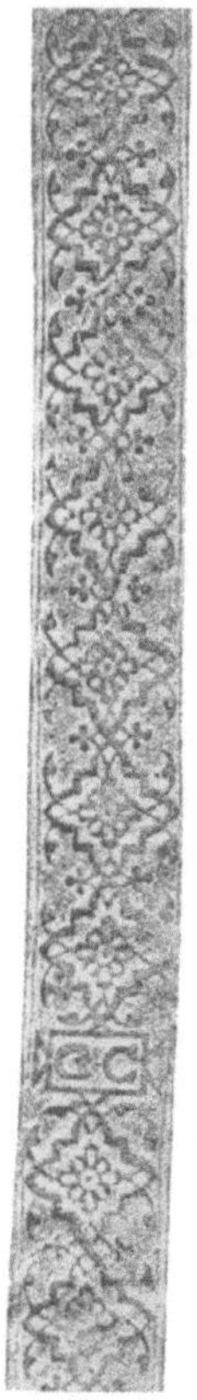

DI. *h* (1)
615

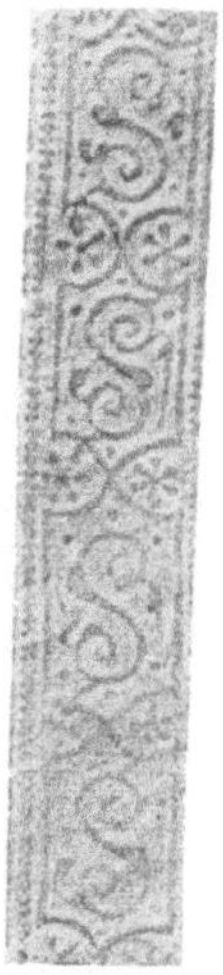

DI. *h* (2)
616

DI. *h* (3)
617

DI. *h* (4)
618

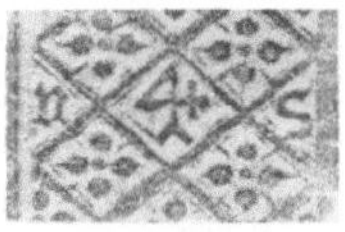

DI. *a* (3)
605

DI. *a* (2)
604

DI. *a* (1)
593

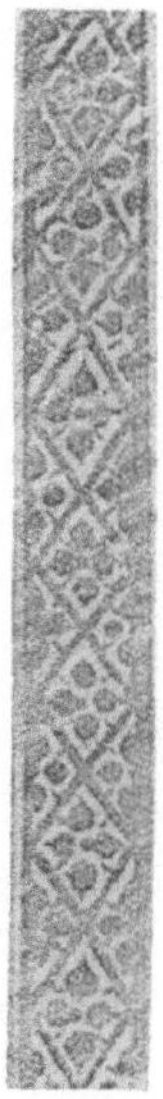

DI. *a* (4)
594

DI. *a* (5)
595

DI. *a* (6)
596

DI. *a* (7)
597

DI. *a* (8)
598

DI. *a* (9)
599

DI. *a* (10)
600

DI. *a* (11)
601

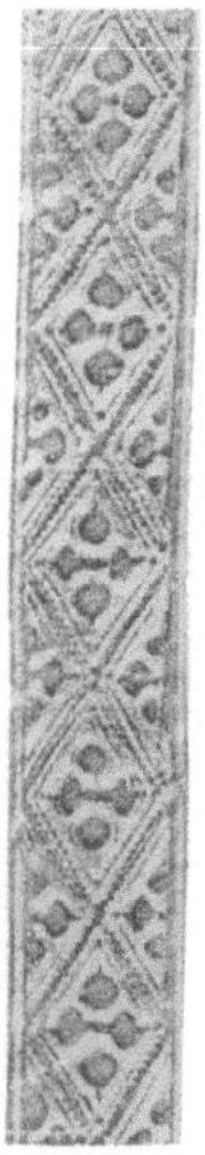

DI. *a* (12)
602

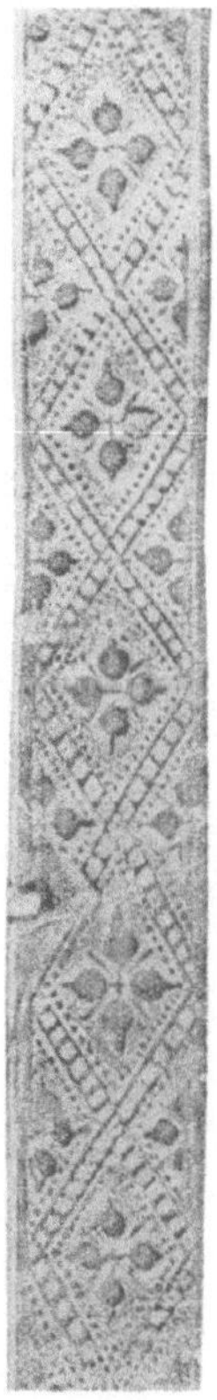

DI. *a* (13)
603

CLASSIFIED ROLLS DI.

PLATE XL

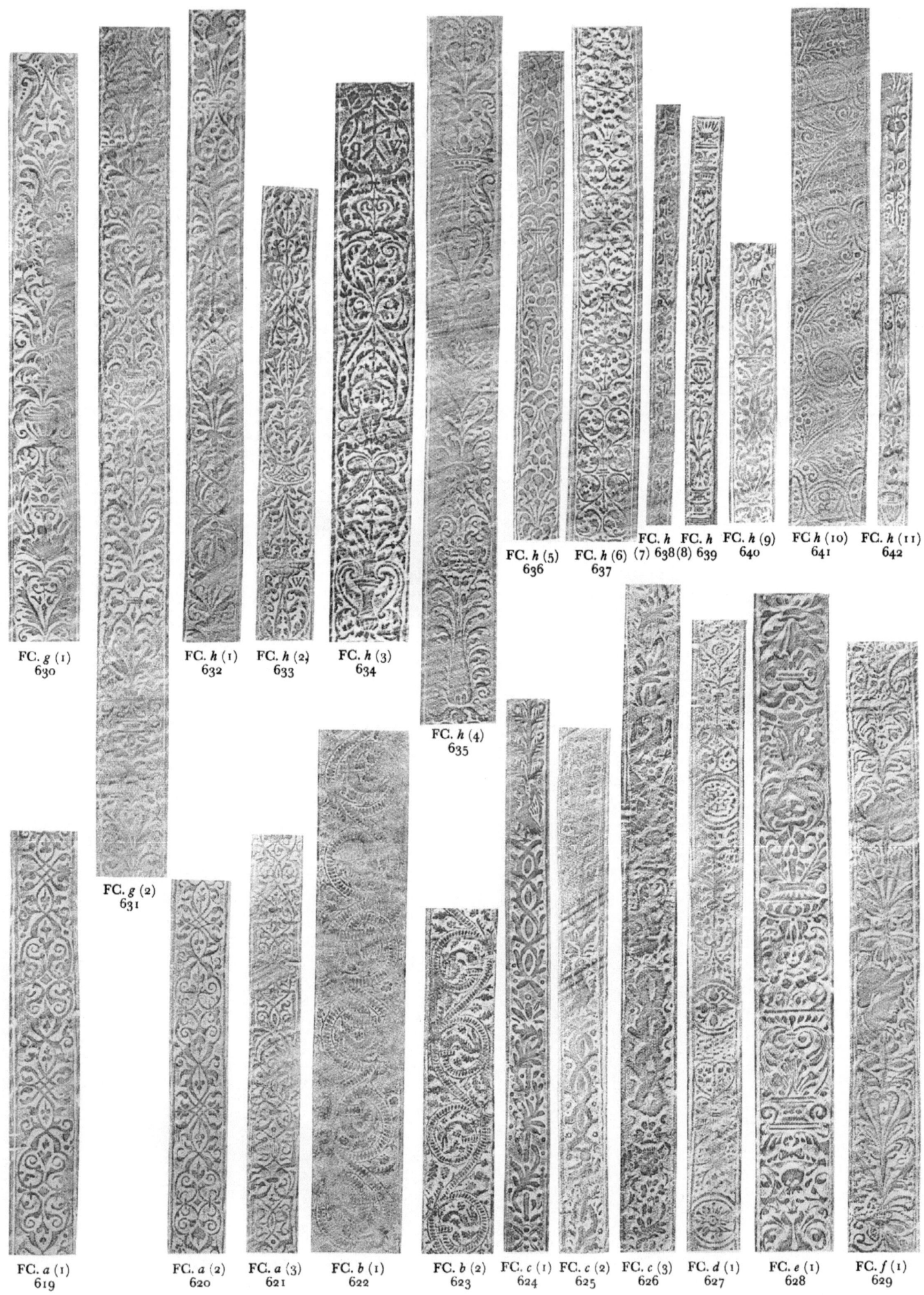

FC. *g* (1)
630

FC. *g* (2)
631

FC. *h* (1)
632

FC. *h* (2)
633

FC. *h* (3)
634

FC. *h* (4)
635

FC. *h* (5)
636

FC. *h* (6)
637

FC. *h*
(7) 638

FC. *h*
(8) 639

FC. *h* (9)
640

FC *h* (10)
641

FC. *h* (11)
642

FC. *a* (1)
619

FC. *a* (2)
620

FC. *a* (3)
621

FC. *b* (1)
622

FC. *b* (2)
623

FC. *c* (1)
624

FC. *c* (2)
625

FC. *c* (3)
626

FC. *d* (1)
627

FC. *e* (1)
628

FC. *f* (1)
629

CLASSIFIED ROLLS FC.

PLATE 41

PLATE XLI

FP. *a* (15) 657   FP. *a* (16) 658   FP. *a* (17) 659   FP. *a* (18) 660   FP. *a* (19) 661   FP. *a* (20) 662   FP. *a* (21) 663   FP. *a* (22) 664   FP. *b* (1) 665   FP. *b* (2) 666   FP. *b* (3) 667   FP. *b* (4) 668   FP. *b* (5) 699   FP. *b* (6) 670

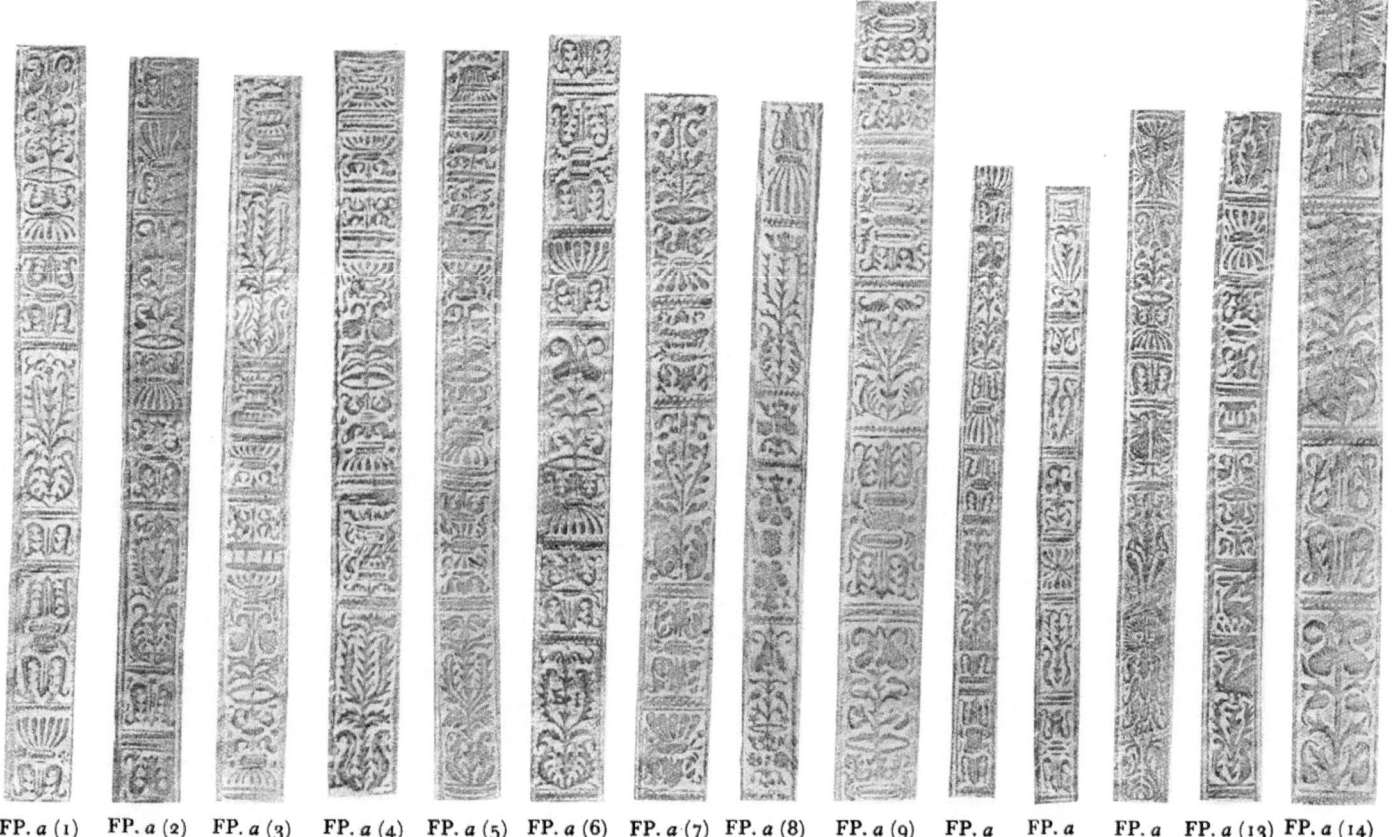

FP. *a* (1) 643   FP. *a* (2) 644   FP. *a* (3) 645   FP. *a* (4) 646   FP. *a* (5) 647   FP. *a* (6) 648   FP. *a* (7) 649   FP. *a* (8) 650   FP. *a* (9) 651   FP. *a* (10) 652   FP. *a* (11) 653   FP. *a* (12) 654   FP. *a* (13) 655   FP. *a* (14) 656

CLASSIFIED ROLLS FP.

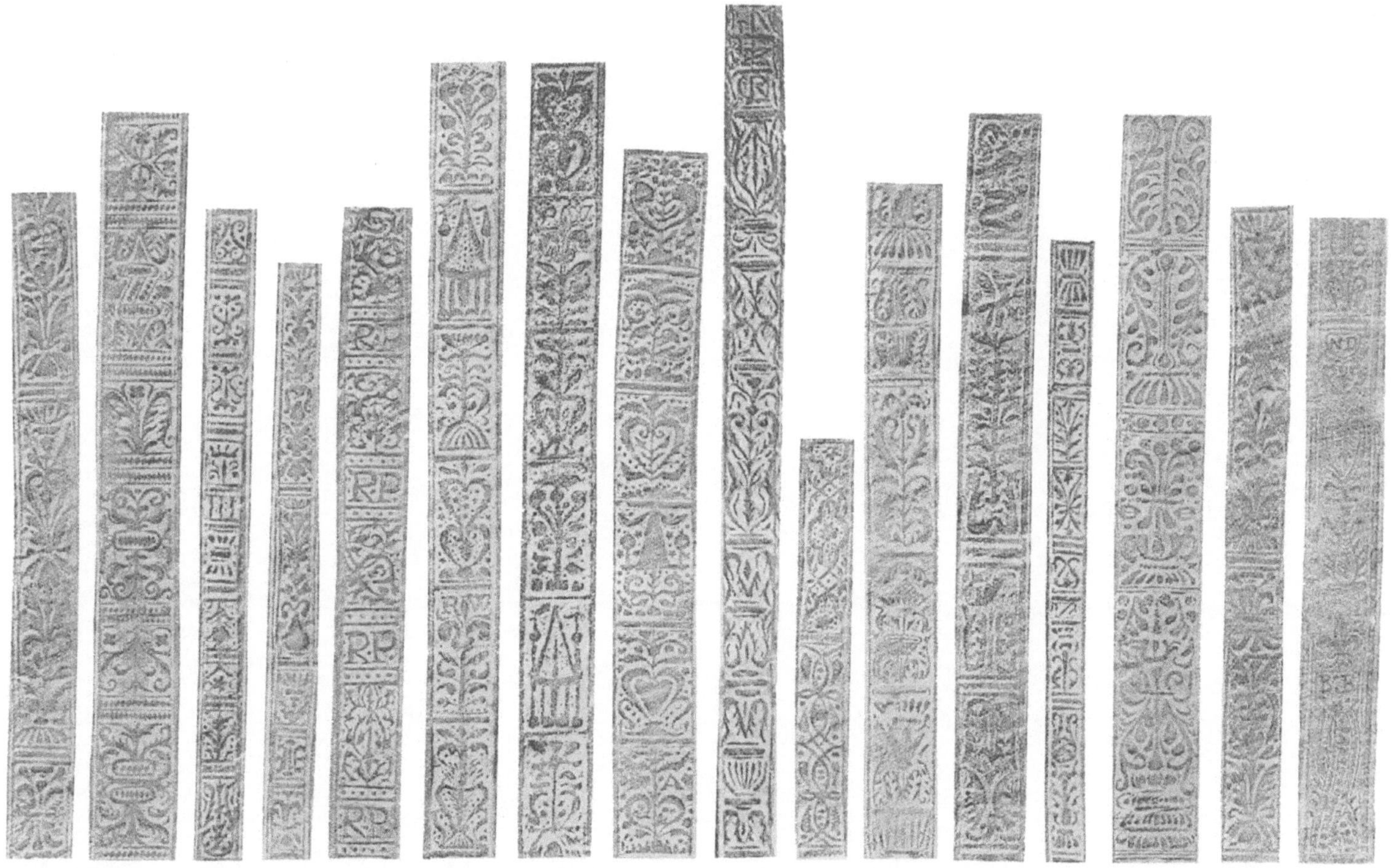

PLATE XLII

FP. *g* (1) 685   FP. *g* (2) 686   FP. *g* (3) 687   FP. *g* (4) 688   FP. *g* (5) 689   FP. *g* (6) 690   FP. *g* (6a) 691   FP. *g* (7) 692   FP. *g* (8) 693   FP. *g* (9) 694   FP. *g* (10) 695   FPl *g* (11) 696   FP. *g* (12) 697   FP. *g* (13) 698   FP. *g* (14) 699   FP. *g* (15) 700

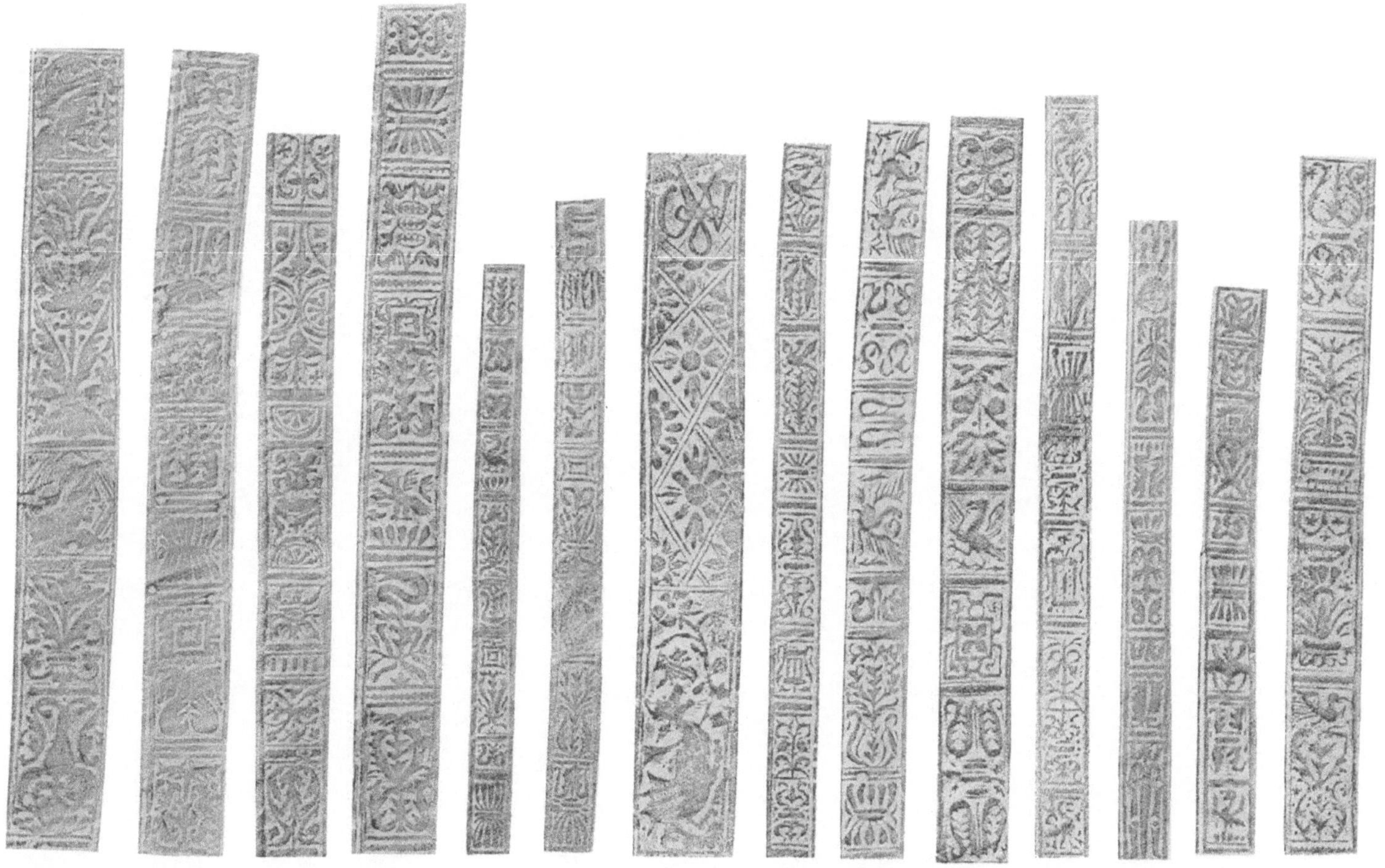

FP. *c* (1) 671   FP. *d* (1) 672   FP. *e* (1) 673   FP. *f* (1) 674   FP. *f* (2) 675   FP. *f* (3) 676   FP. *f* (4) 677   FP. *f* (5) 678   FP. *f* (6) 679   FP. *f* (7) 680   FP. *f* (8) 681   FP. *f* (9) 682   FP. *f* (10) 683   FP. *f* (11) 684

CLASSIFIED ROLLS FP.

PLATE XLIII

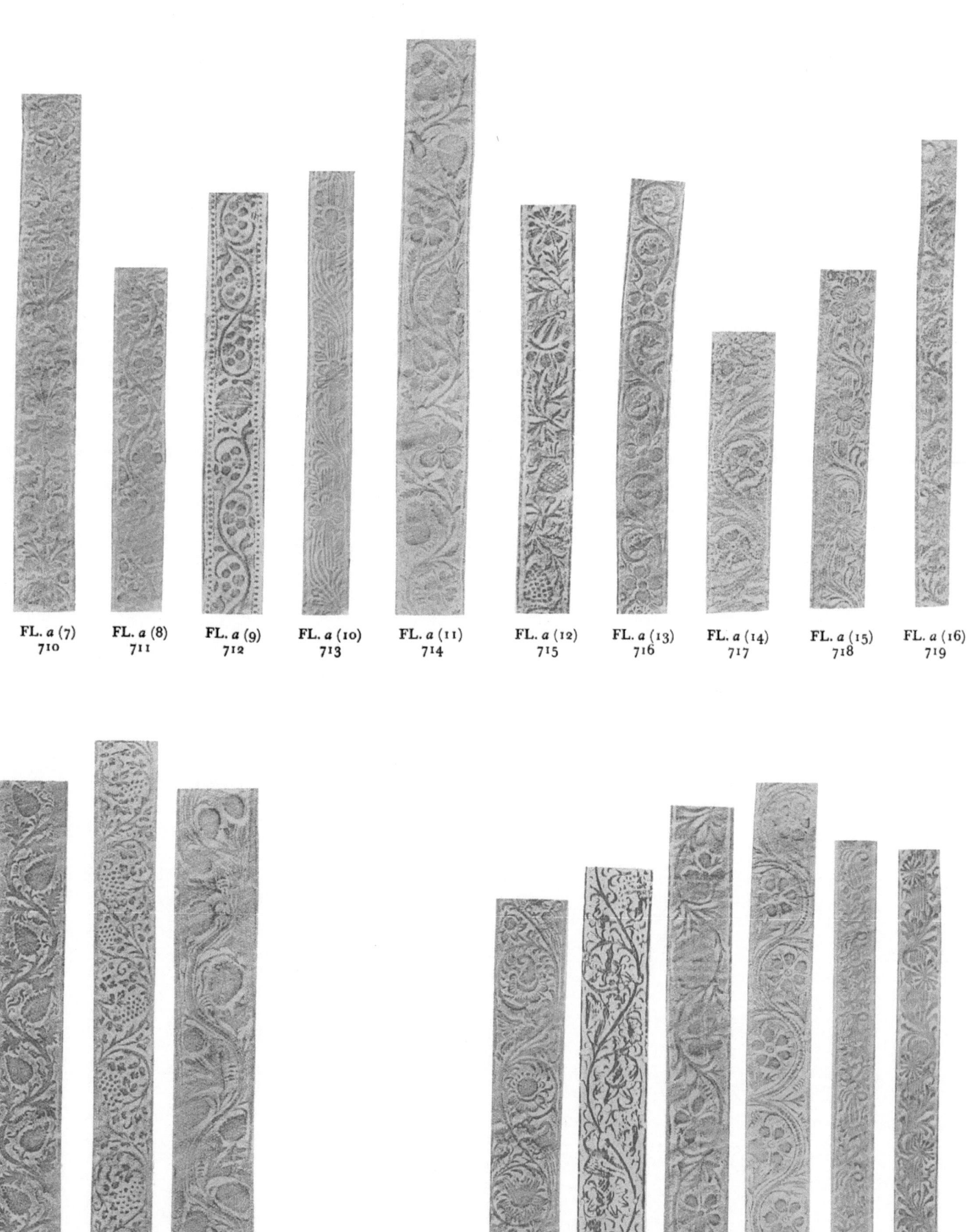

FL. *a* (7)   FL. *a* (8)   FL. *a* (9)   FL. *a* (10)   FL. *a* (11)   FL. *a* (12)   FL. *a* (13)   FL. *a* (14)   FL. *a* (15)   FL. *a* (16)
710   711   712   713   714   715   716   717   718   719

FR. (1)   FR. (2)   FR. (3)   FL. *a* (1)   FL. *a* (2)   FL. *a* (3)   FL. *a* (4)   FL. *a* (5)   FL. *a* (6)
701   702   703   704   705   706   707   708   709

PLATE XLIII

PLATE XLIV

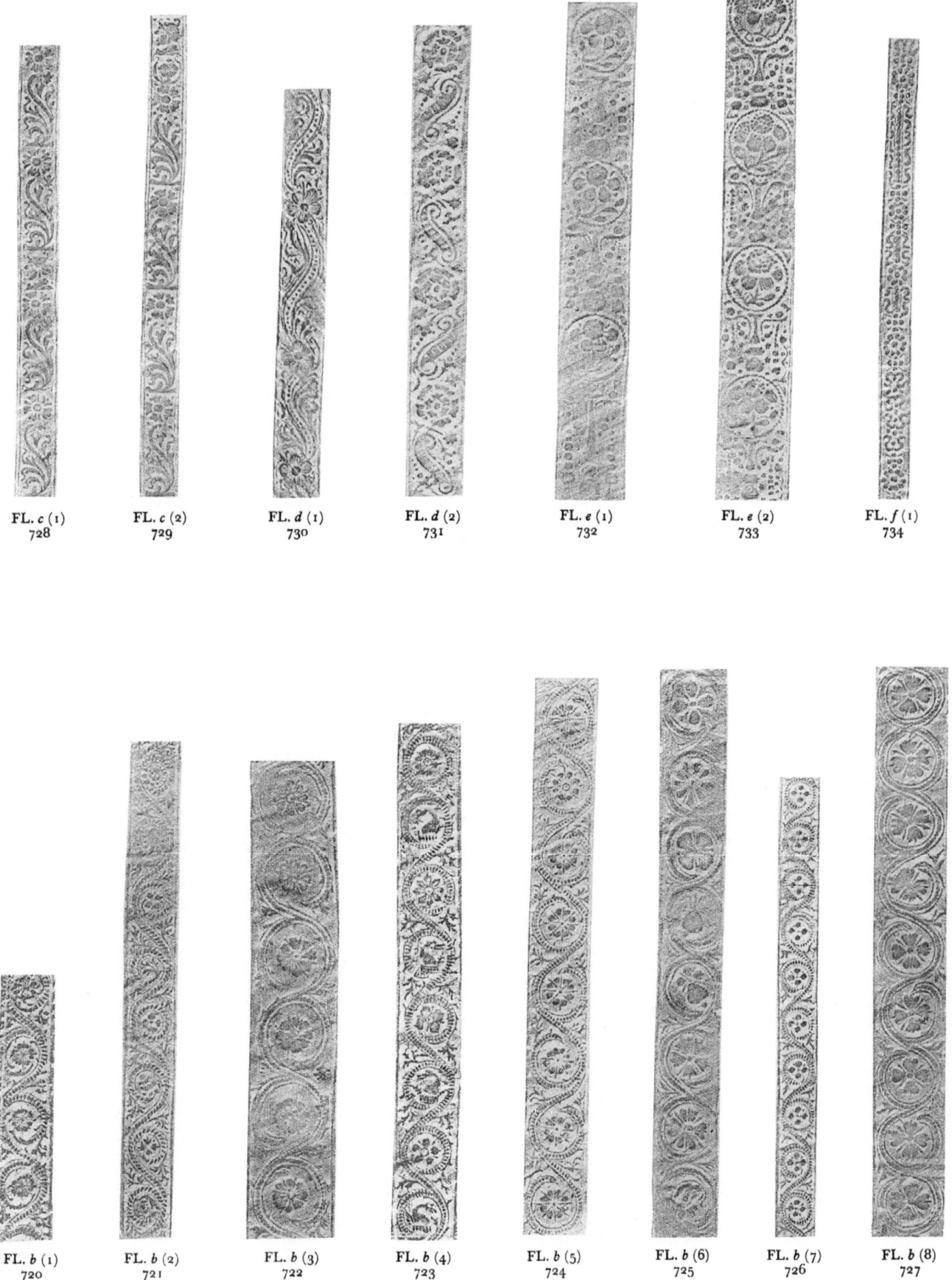

FL. *c* (1)
728

FL. *c* (2)
729

FL. *d* (1)
730

FL. *d* (2)
731

FL. *e* (1)
732

FL. *e* (2)
733

FL. *f* (1)
734

FL. *b* (1)
720

FL. *b* (2)
721

FL. *b* (3)
722

FL. *b* (4)
723

FL. *b* (5)
724

FL. *b* (6)
725

FL. *b* (7)
726

FL. *b* (8)
727

CLASSIFIED ROLLS FL.

PLATE XLV

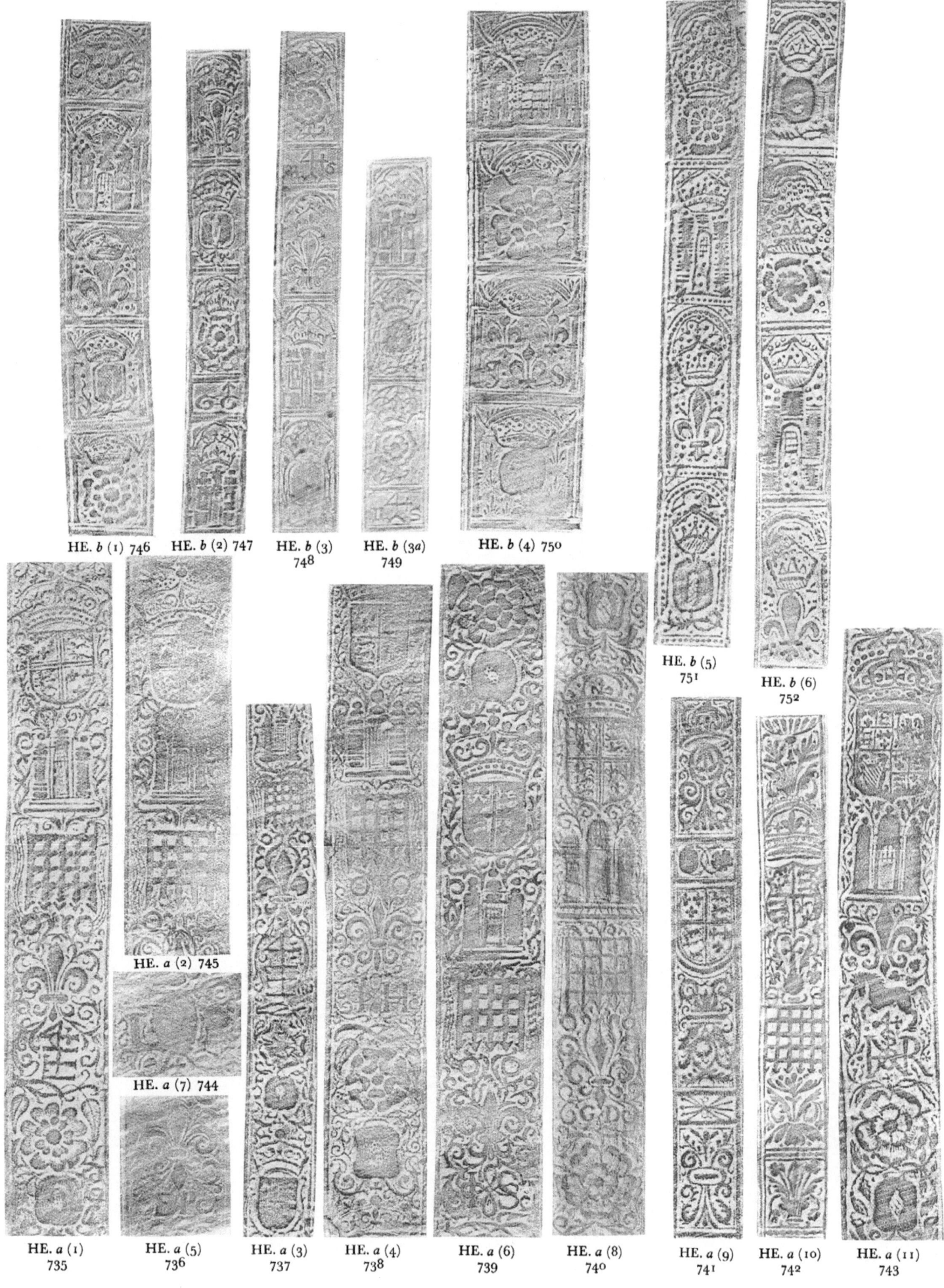

HE. b (1) 746   HE. b (2) 747   HE. b (3)   HE. b (3a)   HE. b (4) 750
748   749

HE. b (5)   HE. b (6)
751   752

HE. a (2) 745

HE. a (7) 744

HE. a (1)   HE. a (5)   HE. a (3)   HE. a (4)   HE. a (6)   HE. a (8)   HE. a (9)   HE. a (10)   HE. a (11)
735   736   737   738   739   740   741   742   743

CLASSIFIED ROLLS HE.

PLATE XIV

PLATE XLVI

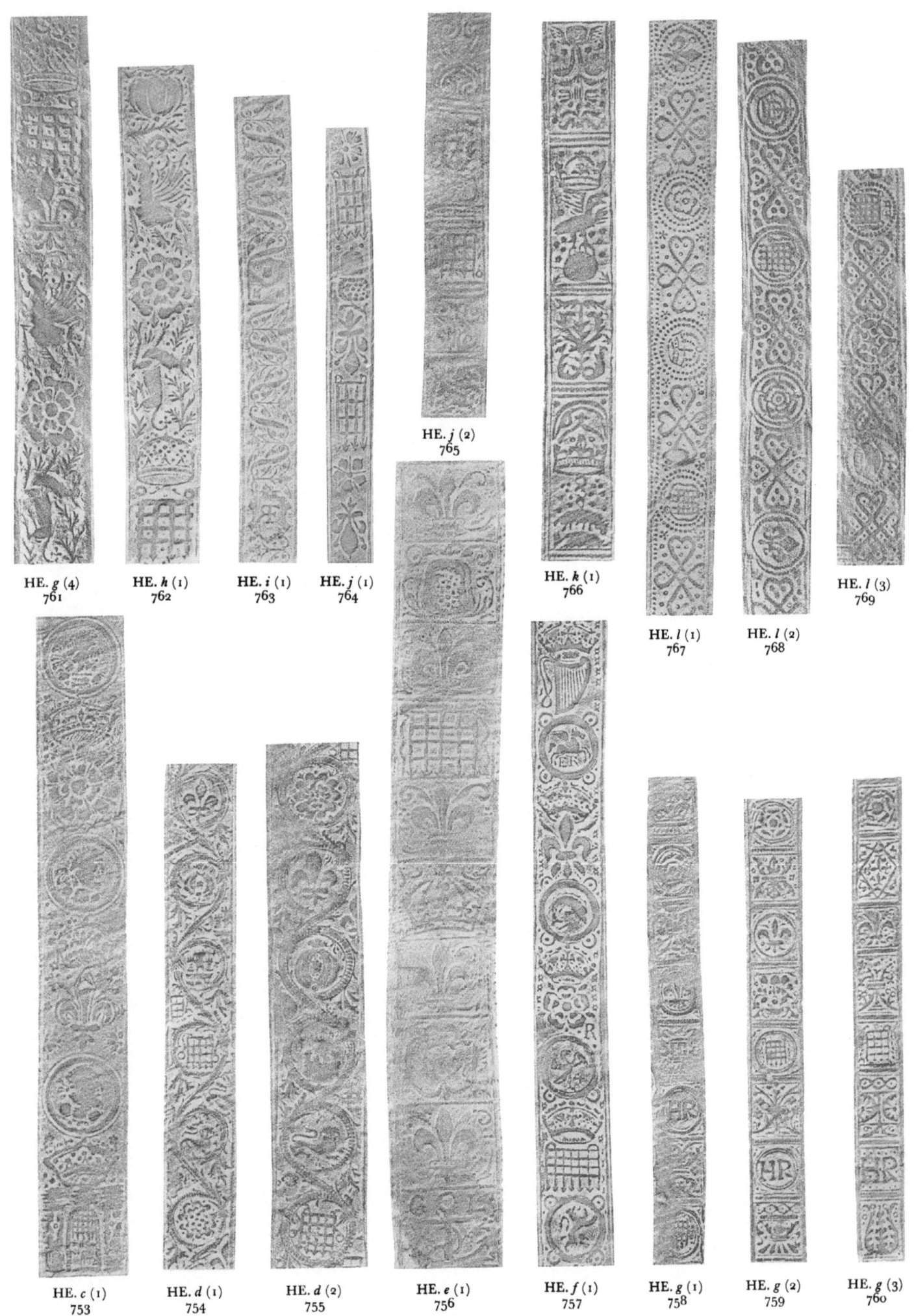

HE. *g* (4)
761

HE. *h* (1)
762

HE. *i* (1)
763

HE. *j* (1)
764

HE. *j* (2)
765

HE. *k* (1)
766

HE. *l* (1)
767

HE. *l* (2)
768

HE. *l* (3)
769

HE. *c* (1)
753

HE. *d* (1)
754

HE. *d* (2)
755

HE. *e* (1)
756

HE. *f* (1)
757

HE. *g* (1)
758

HE. *g* (2)
759

HE. *g* (3)
760

CLASSIFIED ROLLS HE.

PLATE XLVII

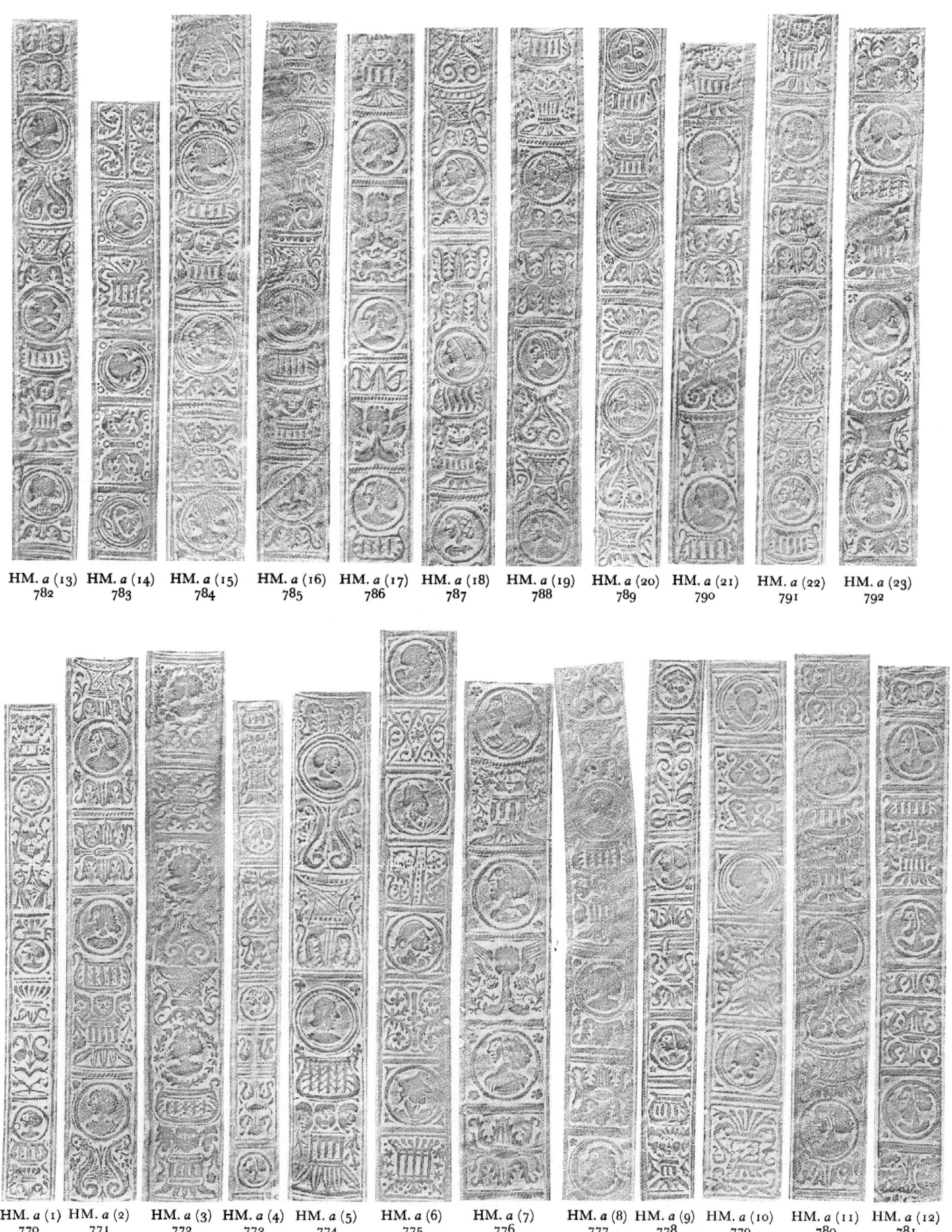

HM. *a* (13) 782   HM. *a* (14) 783   HM. *a* (15) 784   HM. *a* (16) 785   HM. *a* (17) 786   HM. *a* (18) 787   HM. *a* (19) 788   HM. *a* (20) 789   HM. *a* (21) 790   HM. *a* (22) 791   HM. *a* (23) 792

HM. *a* (1) 770   HM. *a* (2) 771   HM. *a* (3) 772   HM. *a* (4) 773   HM. *a* (5) 774   HM. *a* (6) 775   HM. *a* (7) 776   HM. *a* (8) 777   HM. *a* (9) 778   HM. *a* (10) 779   HM. *a* (11) 780   HM. *a* (12) 781

PLATE XLVIII

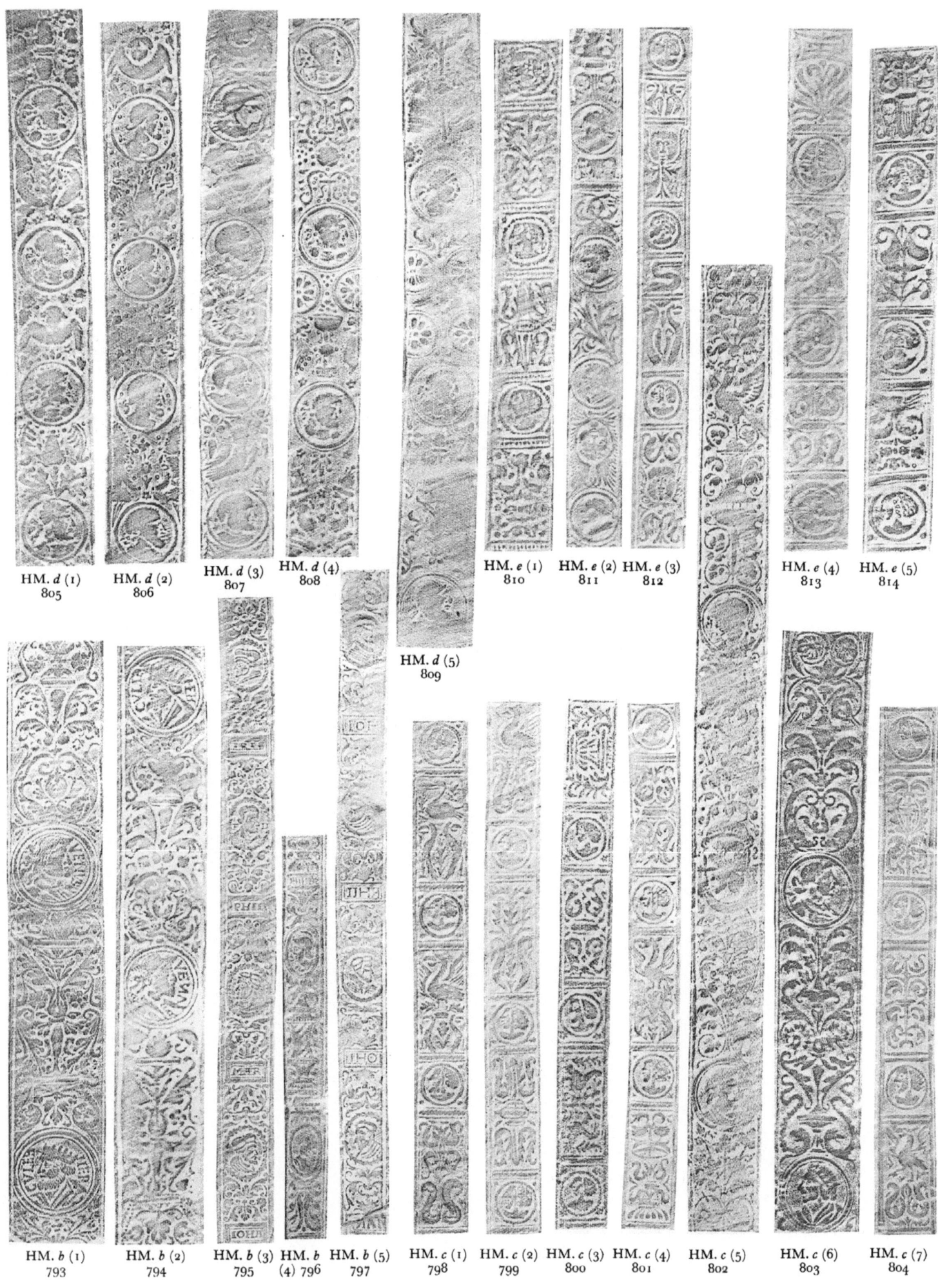

HM. *d* (1)   HM. *d* (2)   HM. *d* (3)   HM. *d* (4)   HM. *e* (1)   HM. *e* (2)   HM. *e* (3)   HM. *e* (4)   HM. *e* (5)
805   806   807   808   810   811   812   813   814

HM. *d* (5)
809

HM. *b* (1)   HM. *b* (2)   HM. *b* (3)   HM. *b* (4) 796   HM. *b* (5)   HM. *c* (1)   HM. *c* (2)   HM. *c* (3)   HM. *c* (4)   HM. *c* (5)   HM. *c* (6)   HM. *c* (7)
793   794   795   797   798   799   800   801   802   803   804

CLASSIFIED ROLLS HM.

PLATE XLVIII

PLATE XLIX

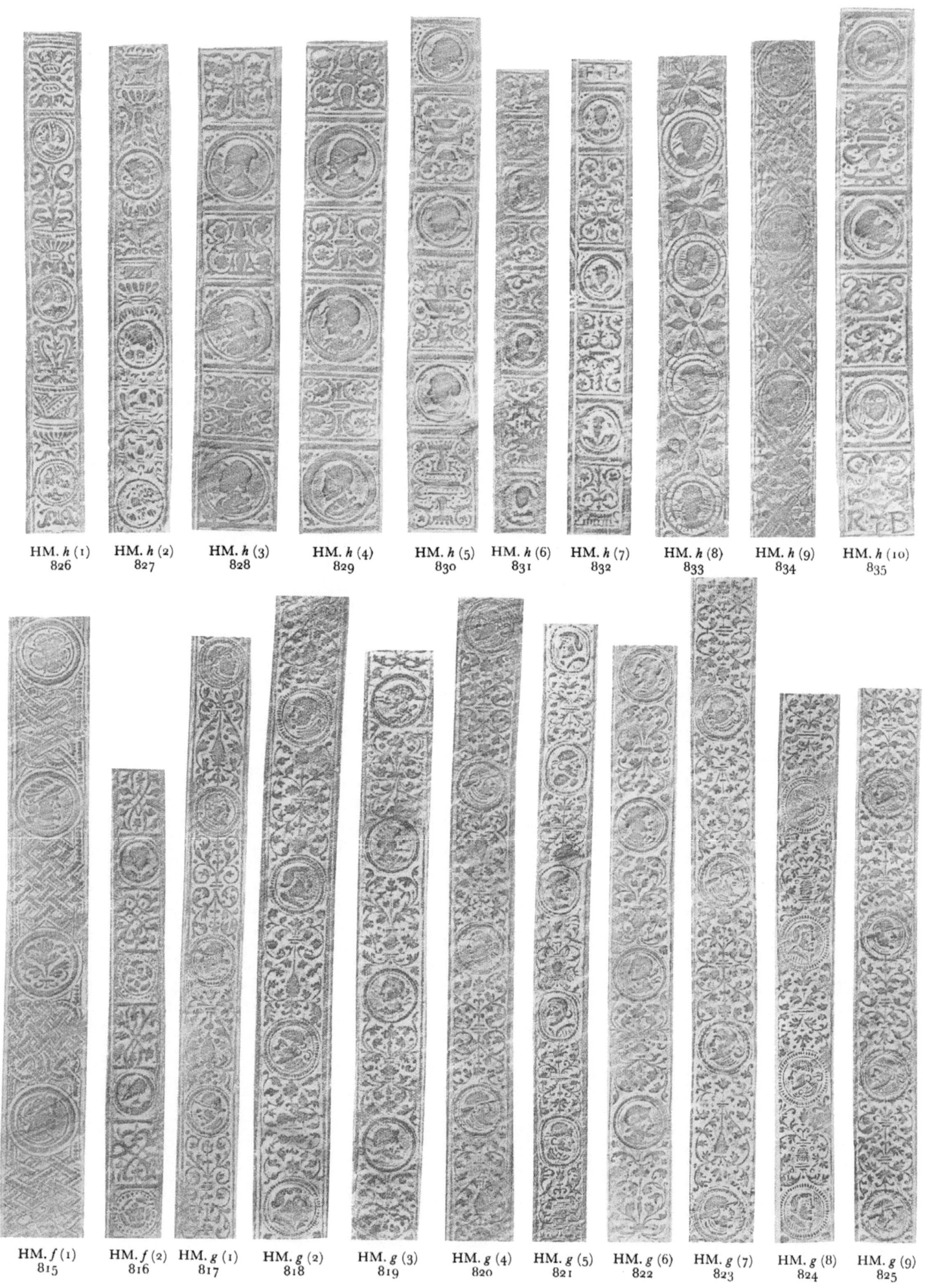

HM. *h* (1)
826

HM. *h* (2)
827

HM. *h* (3)
828

HM. *h* (4)
829

HM. *h* (5)
830

HM. *h* (6)
831

HM. *h* (7)
832

HM. *h* (8)
833

HM. *h* (9)
834

HM. *h* (10)
835

HM. *f* (1)
815

HM. *f* (2)
816

HM. *g* (1)
817

HM. *g* (2)
818

HM. *g* (3)
819

HM. *g* (4)
820

HM. *g* (5)
821

HM. *g* (6)
822

HM. *g* (7)
823

HM. *g* (8)
824

HM. *g* (9)
825

CLASSIFIED ROLLS HM.

PLATE L

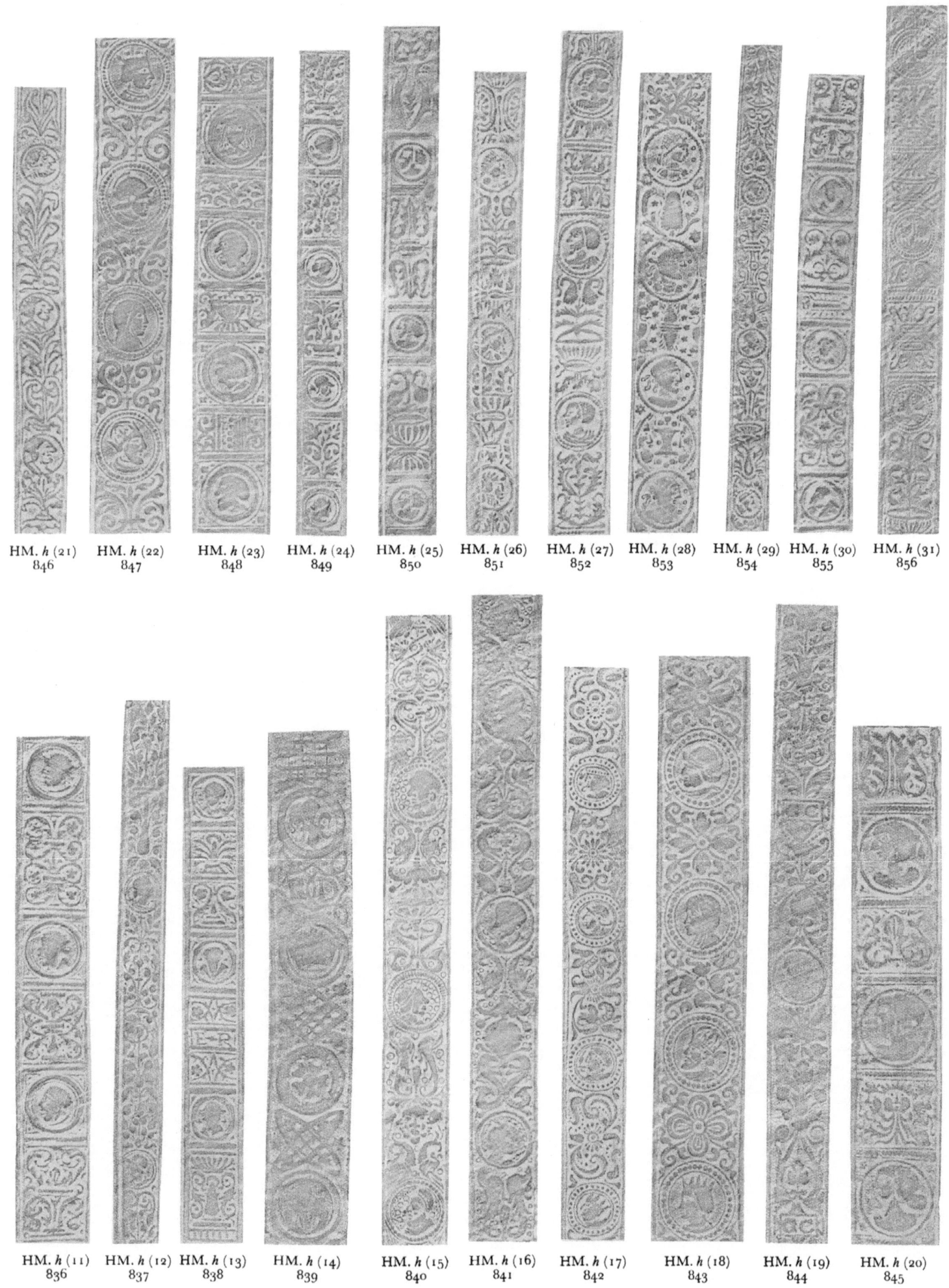

HM. *h* (21)
846

HM. *h* (22)
847

HM. *h* (23)
848

HM. *h* (24)
849

HM. *h* (25)
850

HM. *h* (26)
851

HM. *h* (27)
852

HM. *h* (28)
853

HM. *h* (29)
854

HM. *h* (30)
855

HM. *h* (31)
856

HM. *h* (11)
836

HM. *h* (12)
837

HM. *h* (13)
838

HM. *h* (14)
839

HM. *h* (15)
840

HM. *h* (16)
841

HM. *h* (17)
842

HM. *h* (18)
843

HM. *h* (19)
844

HM. *h* (20)
845

CLASSIFIED ROLLS HM.

PLATE LI

MW. d (5) 867   MW. d (6) 868   MW. d (7) 869   MW. d (8) 870   MW. d (9) 871   MW. d (10) 872   MW. d (11) 873   MW. d (12) 874   MW. d (13) 875   MW. d (14) 876   MW. d (15) 877

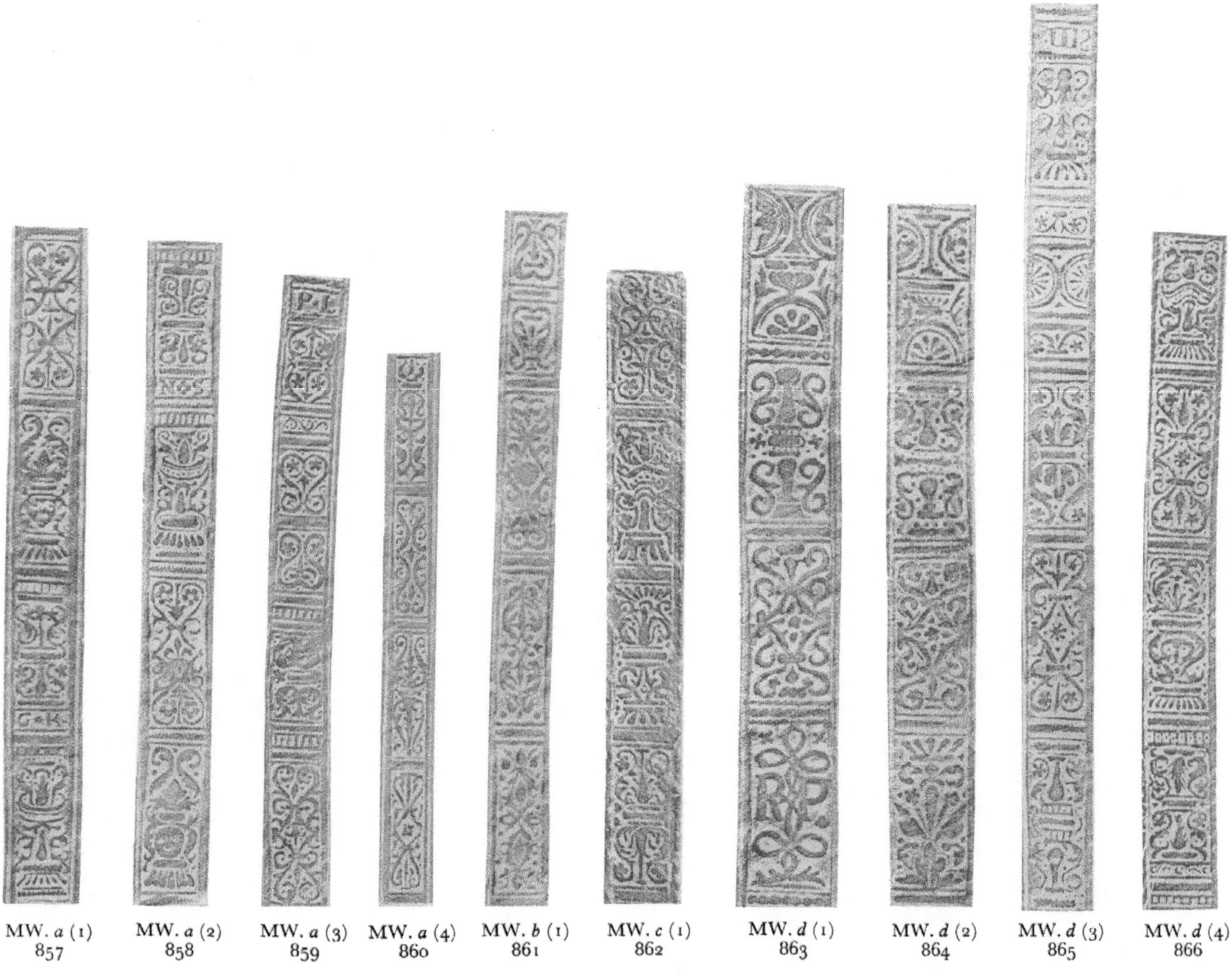

MW. a (1) 857   MW. a (2) 858   MW. a (3) 859   MW. a (4) 860   MW. b (1) 861   MW. c (1) 862   MW. d (1) 863   MW. d (2) 864   MW. d (3) 865   MW. d (4) 866

CLASSIFIED ROLLS MW.

PLATE LII

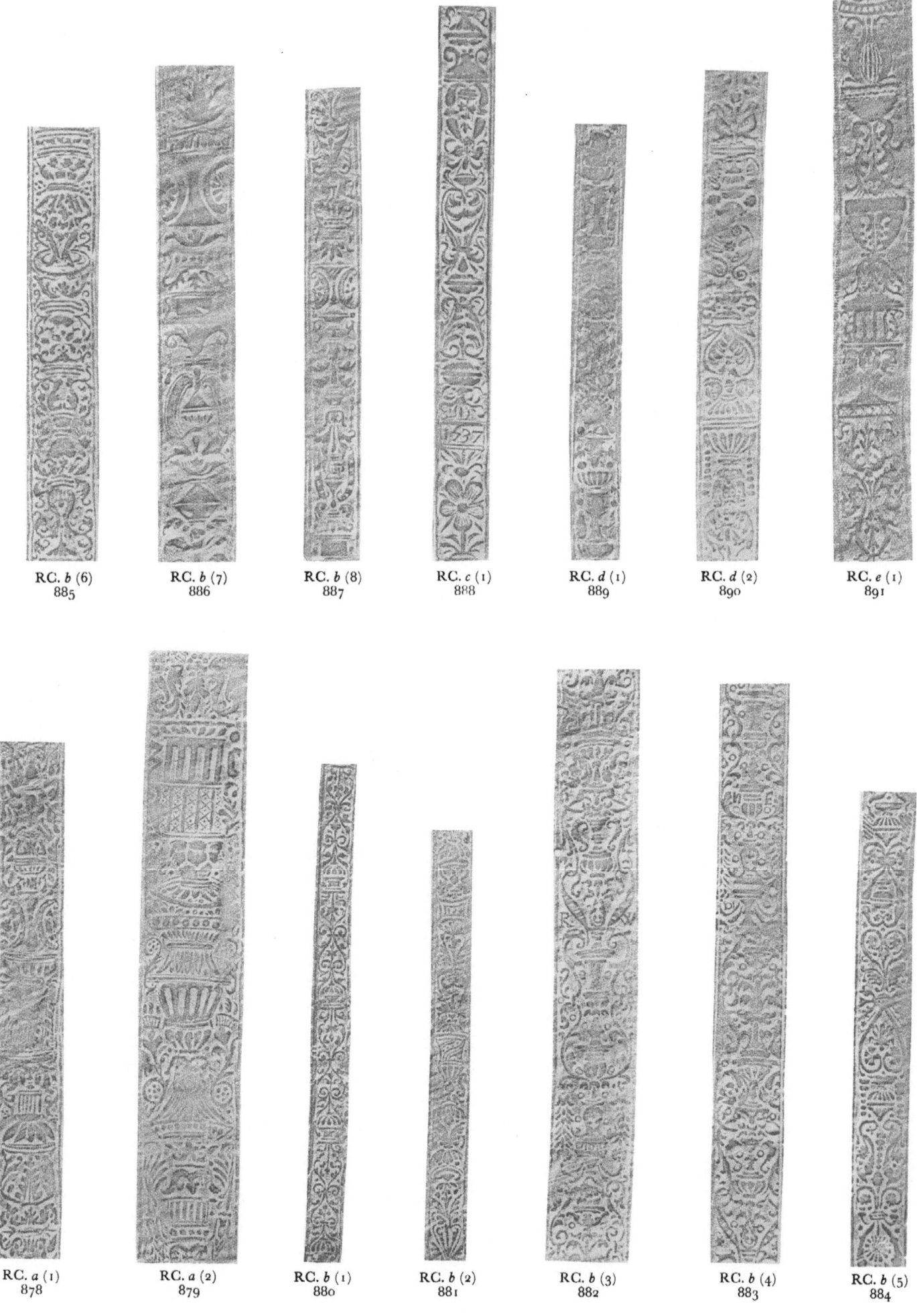

RC. *b* (6)
885

RC. *b* (7)
886

RC. *b* (8)
887

RC. *c* (1)
888

RC. *d* (1)
889

RC. *d* (2)
890

RC. *e* (1)
891

RC. *a* (1)
878

RC. *a* (2)
879

RC. *b* (1)
880

RC. *b* (2)
881

RC. *b* (3)
882

RC. *b* (4)
883

RC. *b* (5)
884

CLASSIFIED ROLLS RC.

PLATE LIII

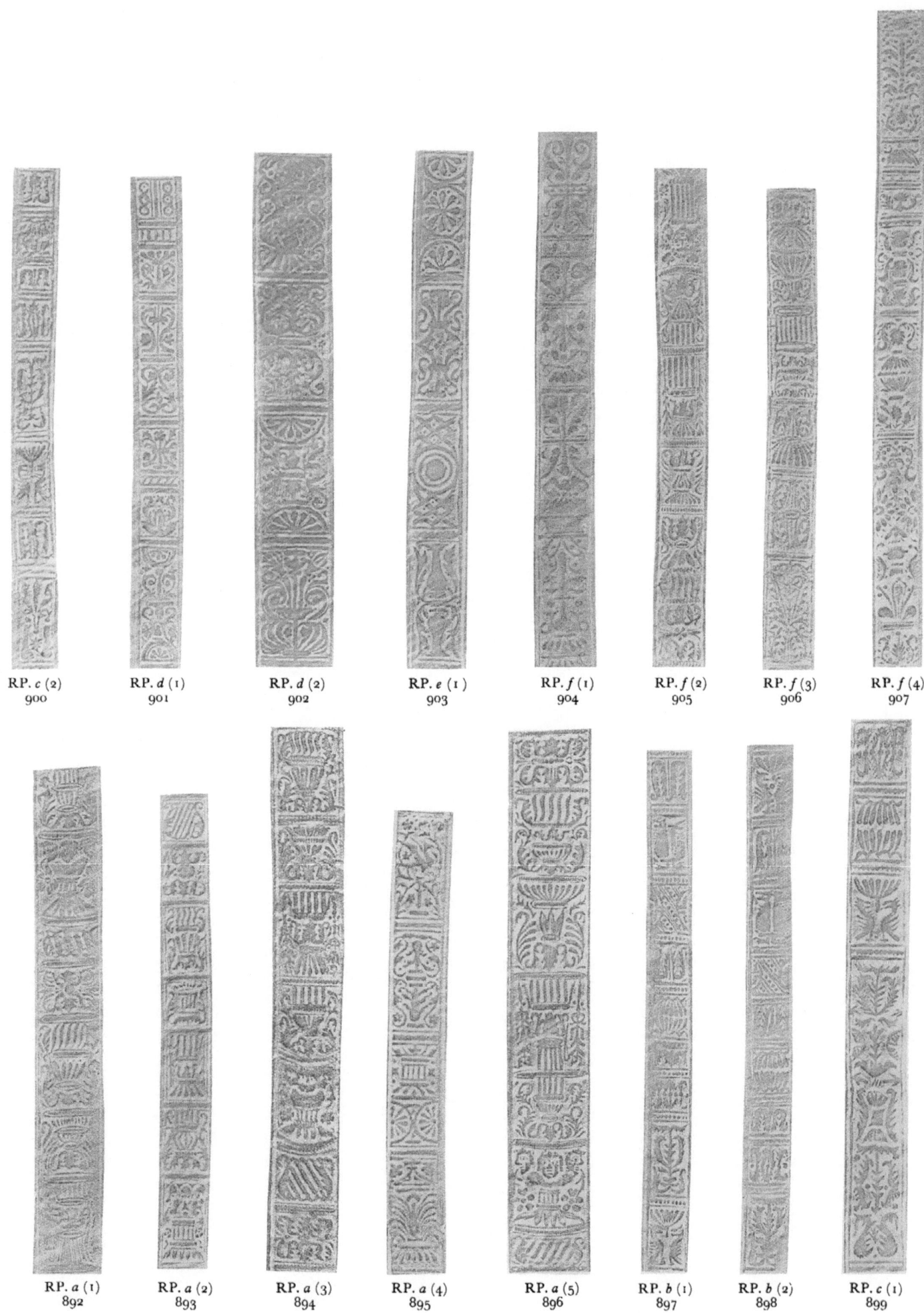

RP. c (2)
900

RP. d (1)
901

RP. d (2)
902

RP. e (1)
903

RP. f (1)
904

RP. f (2)
905

RP. f (3)
906

RP. f (4)
907

RP. a (1)
892

RP. a (2)
893

RP. a (3)
894

RP. a (4)
895

RP. a (5)
896

RP. b (1)
897

RP. b (2)
898

RP. c (1)
899

CLASSIFIED ROLLS RP.

PLATE LIII

PLATE LIV

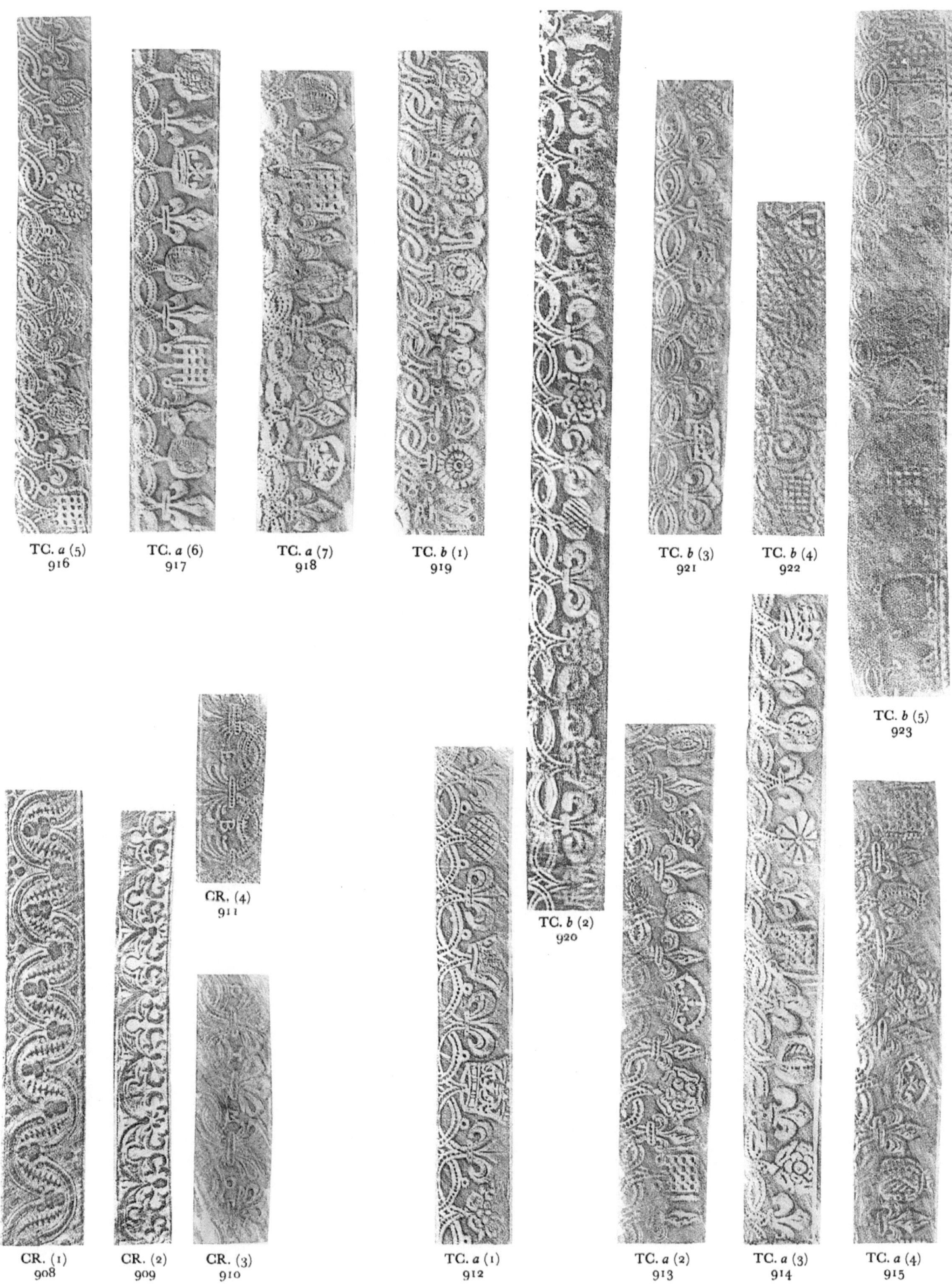

TC. *a* (5)
916

TC. *a* (6)
917

TC. *a* (7)
918

TC. *b* (1)
919

TC. *b* (3)
921

TC. *b* (4)
922

TC. *b* (5)
923

CR. (4)
911

TC. *b* (2)
920

CR. (1)
908

CR. (2)
909

CR. (3)
910

TC. *a* (1)
912

TC. *a* (2)
913

TC. *a* (3)
914

TC. *a* (4)
915

CLASSIFIED ROLLS CR., TC.

PLATE LV

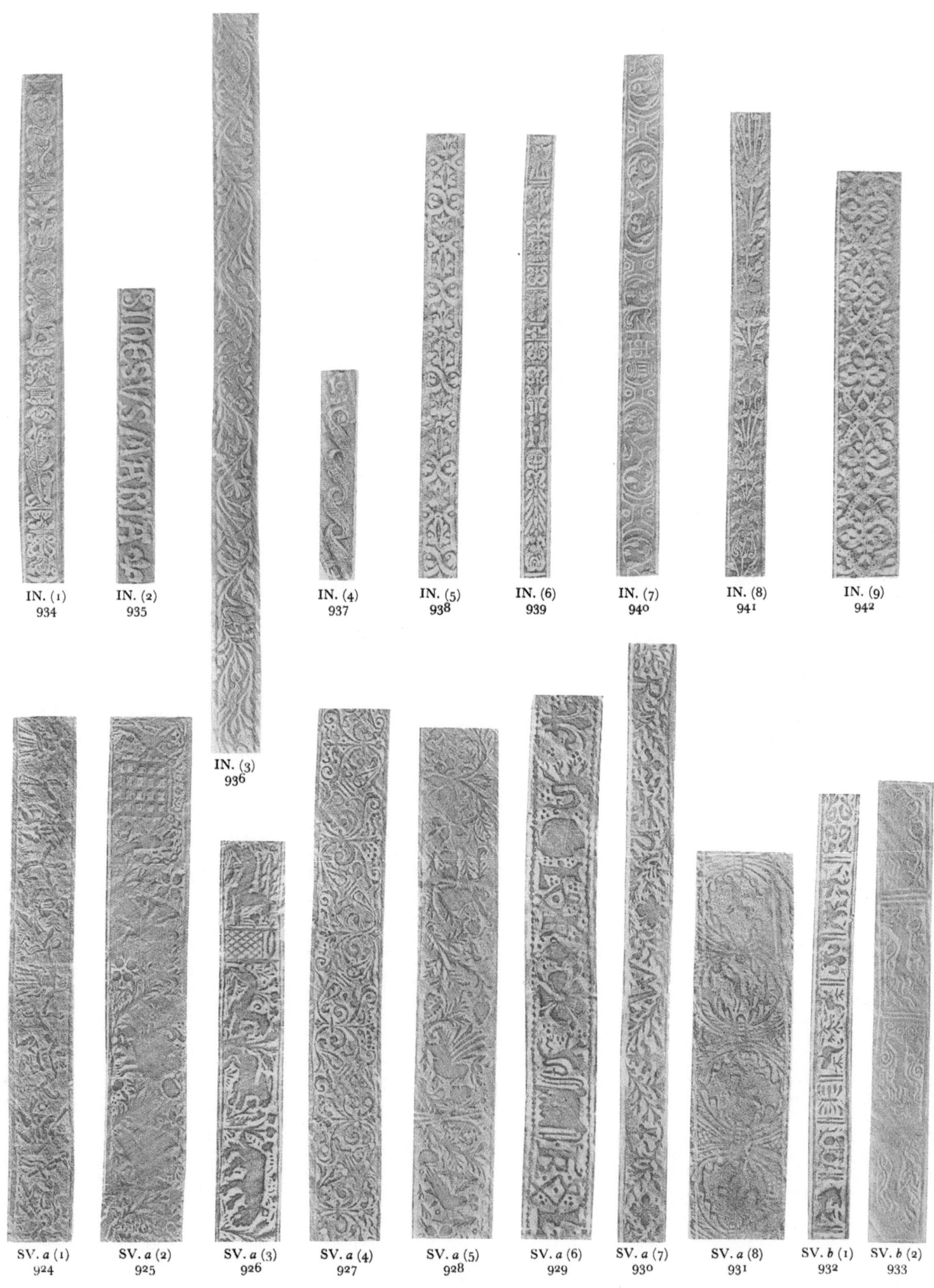

IN. (1)
934

IN. (2)
935

IN. (3)
936

IN. (4)
937

IN. (5)
938

IN. (6)
939

IN. (7)
940

IN. (8)
941

IN. (9)
942

SV. *a* (1)
924

SV. *a* (2)
925

SV. *a* (3)
926

SV. *a* (4)
927

SV. *a* (5)
928

SV. *a* (6)
929

SV. *a* (7)
930

SV. *a* (8)
931

SV. *b* (1)
932

SV. *b* (2)
933

CLASSIFIED ROLLS SV., IN.

PLATE LVI

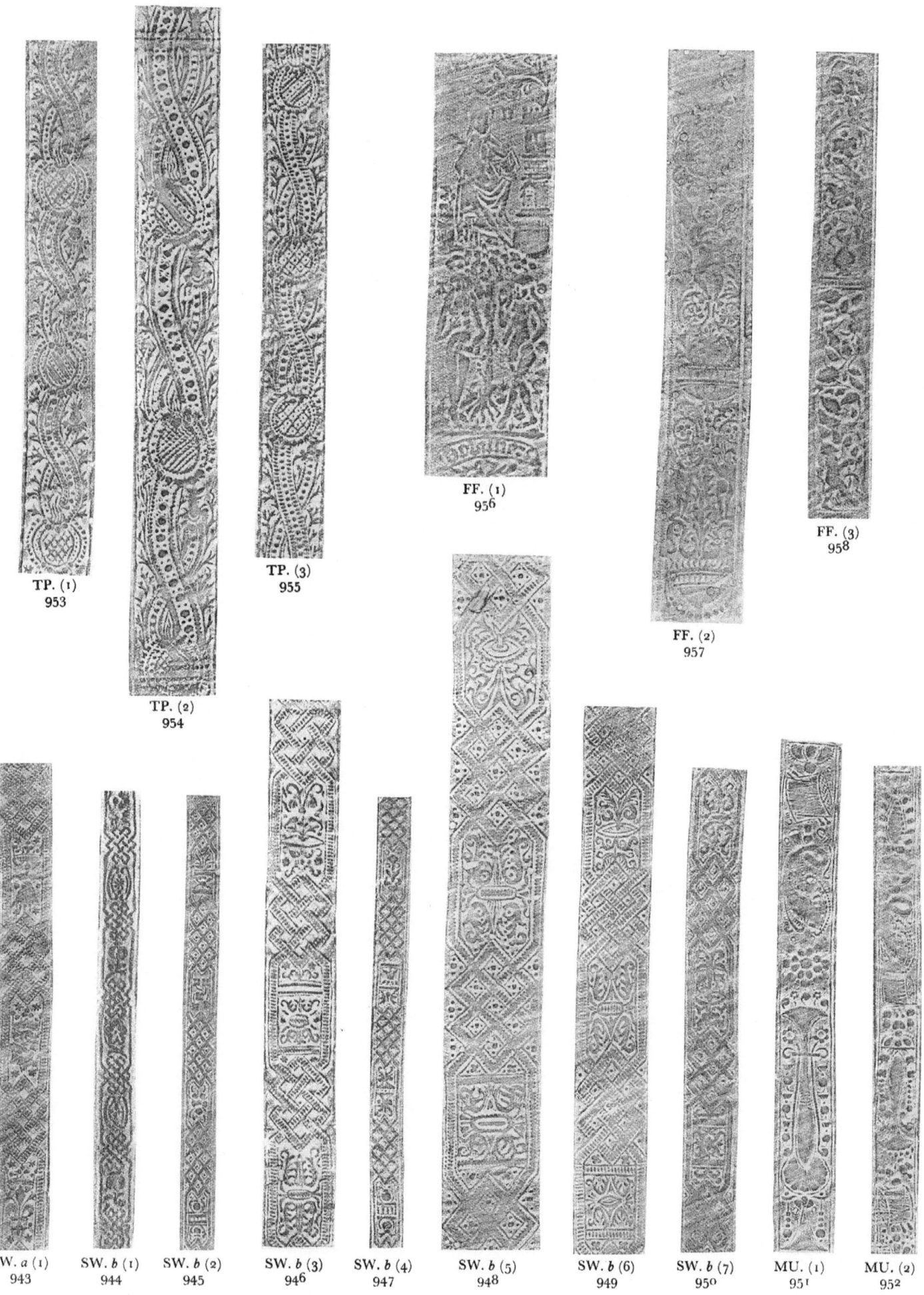

TP. (1)
953

TP. (2)
954

TP. (3)
955

FF. (1)
956

FF. (2)
957

FF. (3)
958

SW. a (1)
943

SW. b (1)
944

SW. b (2)
945

SW. b (3)
946

SW. b (4)
947

SW. b (5)
948

SW. b (6)
949

SW. b (7)
950

MU. (1)
951

MU. (2)
952

CLASSIFIED ROLLS SW., MU., TP., FF.

PLATE LVII

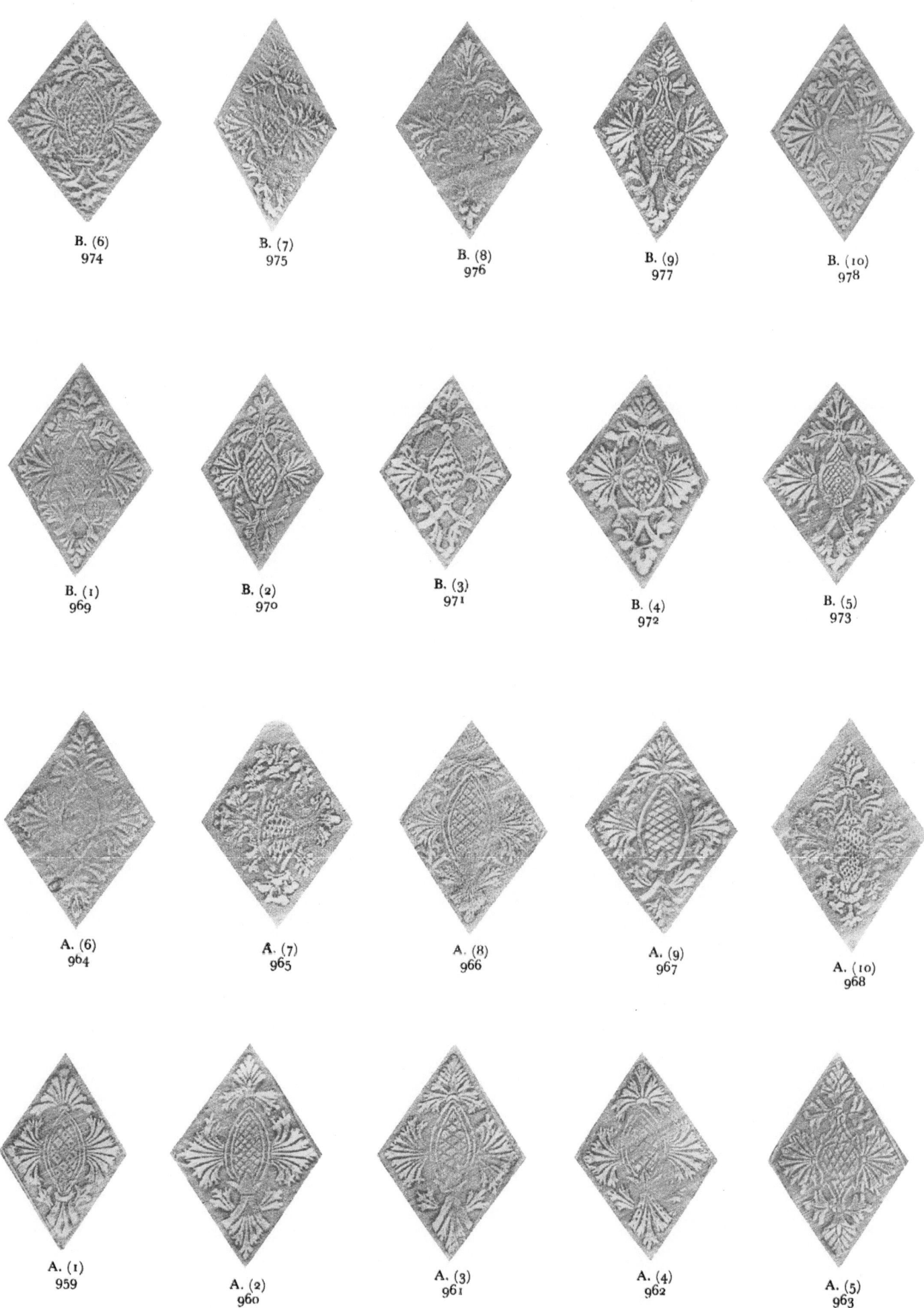

B. (6)
974

B. (7)
975

B. (8)
976

B. (9)
977

B. (10)
978

B. (1)
969

B. (2)
970

B. (3)
971

B. (4)
972

B. (5)
973

A. (6)
964

A. (7)
965

A. (8)
966

A. (9)
967

A. (10)
968

A. (1)
959

A. (2)
960

A. (3)
961

A. (4)
962

A. (5)
963

CLASSIFIED ORNAMENTS A, B

PLATE XVII.

PLATE LVIII

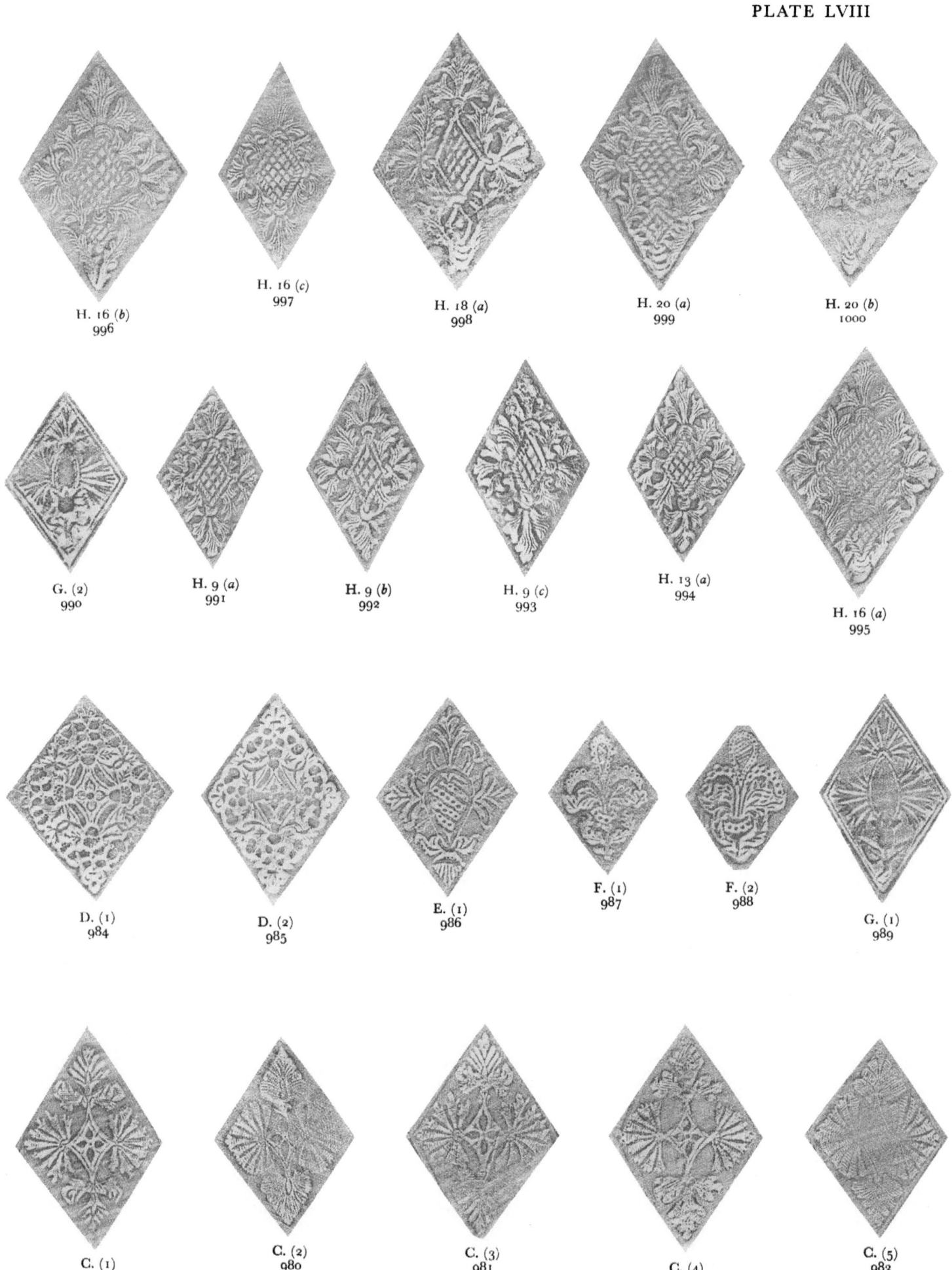

H. 16 (b)
996

H. 16 (c)
997

H. 18 (a)
998

H. 20 (a)
999

H. 20 (b)
1000

G. (2)
990

H. 9 (a)
991

H. 9 (b)
992

H. 9 (c)
993

H. 13 (a)
994

H. 16 (a)
995

D. (1)
984

D. (2)
985

E. (1)
986

F. (1)
987

F. (2)
988

G. (1)
989

C. (1)
979

C. (2)
980

C. (3)
981

C. (4)
982

C. (5)
983

CLASSIFIED ORNAMENTS C–H.

PLATE XXII.

PLATE LIX

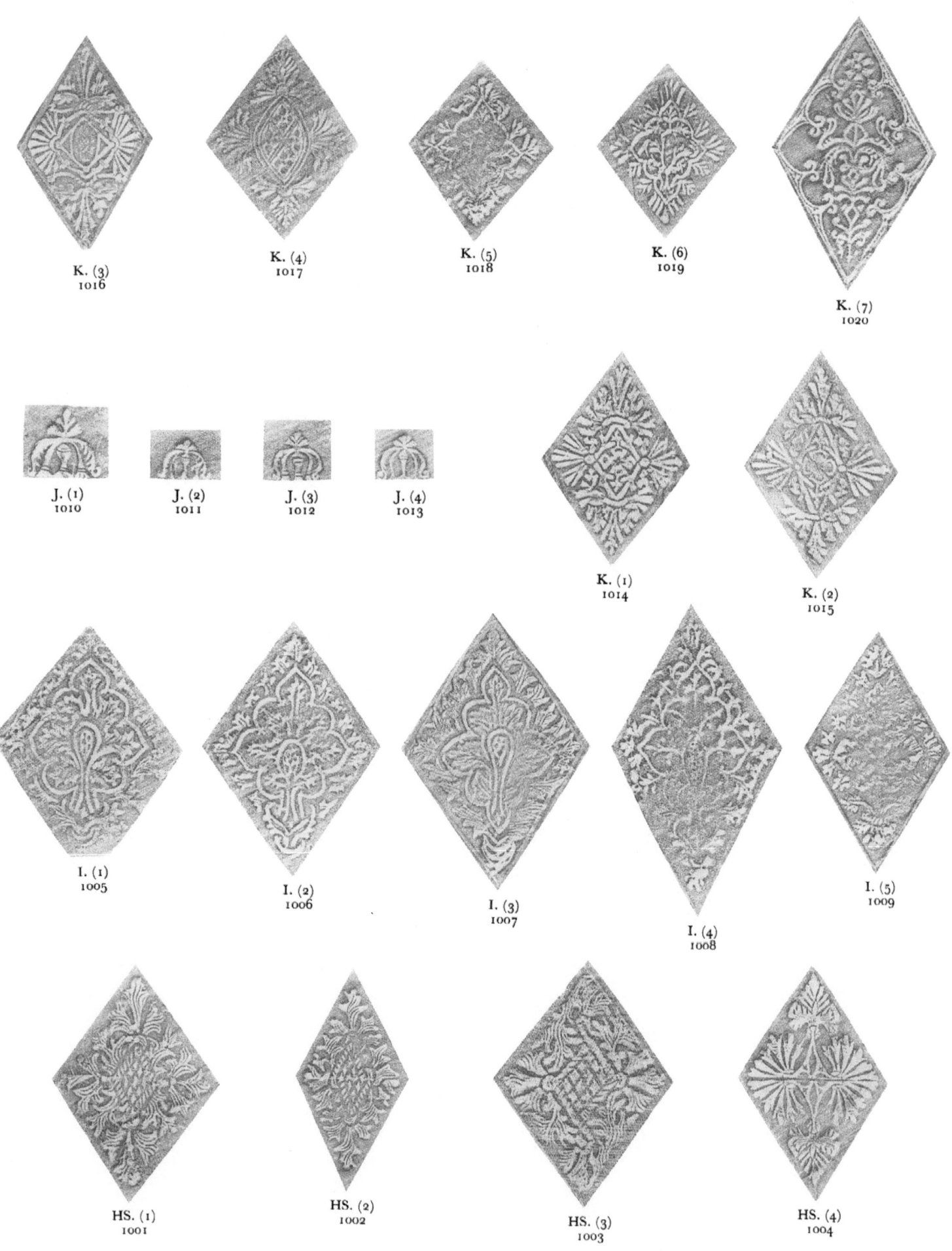

K. (3)
1016

K. (4)
1017

K. (5)
1018

K. (6)
1019

K. (7)
1020

J. (1)
1010

J. (2)
1011

J. (3)
1012

J. (4)
1013

K. (1)
1014

K. (2)
1015

I. (1)
1005

I. (2)
1006

I. (3)
1007

I. (4)
1008

I. (5)
1009

HS. (1)
1001

HS. (2)
1002

HS. (3)
1003

HS. (4)
1004

CLASSIFIED ORNAMENTS HS–K.

PLATE LX

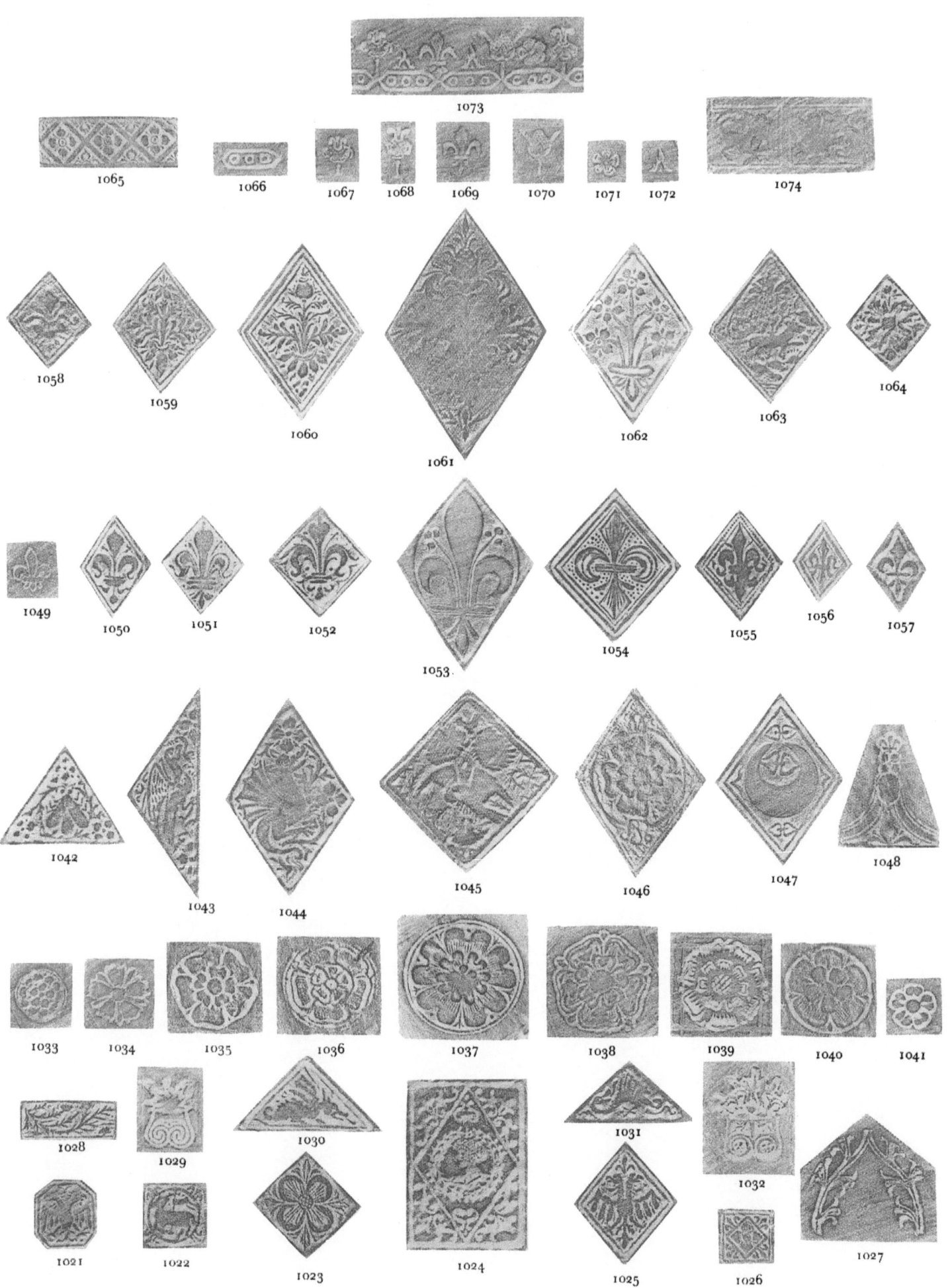

STAMPS USED WITH ROLLS AND ORNAMENTS

PLATE LXI

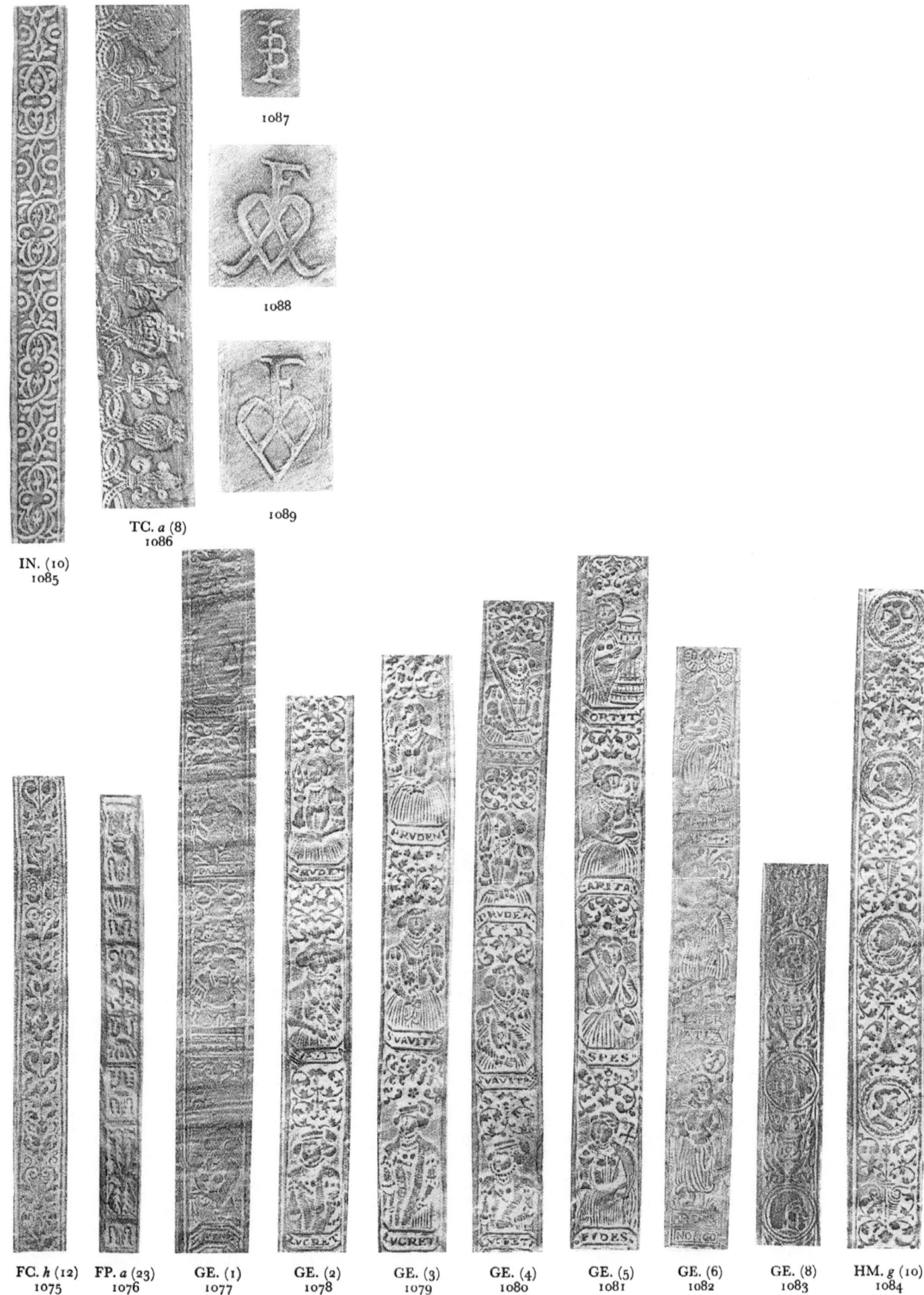

1087

1088

1089

TC. *a* (8)
1086

IN. (10)
1085

FC. *h* (12)
1075

FP. *a* (23)
1076

GE. (1)
1077

GE. (2)
1078

GE. (3)
1079

GE. (4)
1080

GE. (5)
1081

GE. (6)
1082

GE. (8)
1083

HM. *g* (10)
1084

SUPPLEMENTARY